For a classic friend
Mark

With Our Hands

The Story of Carpenters in Massachusetts

With Our Hands

The Story of Carpenters in Massachusetts

Mark Erlich

With the Research Assistance
of David Goldberg

Temple University Press
Philadelphia

Temple University Press, Philadelphia 19122
Copyright © 1986 by Temple University
All rights reserved
Published 1986
Printed in the United States of America

The paper used in this publication meets the minimum
requirements of American National Standard for Information Sciences—Permanence of Paper for Printed
Library Materials, ANSI Z39.48-1984

Library of Congress Cataloging-in-Publication Data

Erlich, Mark, 1949–
  With our hands.

  Includes index.
  1. Carpentry—Massachusetts—History. 2. Building,
Wooden—Massachusetts—History. 3. Trade-unions—
Carpenters—Massachusetts—History. I. Title.
TH5604.E75 1986   331.12′994′09744   86-5992
ISBN 0-87722-433-1 (alk. paper)

Photo Credits

The photographs in the picture galleries are used with
the permission and by the courtesy of the following per-
sons and institutions:

FRED ERNEST   The photo in "Working Lives/Union Life:
The Early Years" was retouched by David DuBusc.
Ernest's father Karl, a bricklayer who appears in the
group, gave the photo to his son.

JOHN GREENLAND   Past president Carpenters Local 40

BPL	Boston Public Library
BRA	Boston Redevelopment Authority
CHC	Cambridge Historical Commission
CL 108	Carpenters Local 108, Springfield
GMMA	George Meany Memorial Archives, Washington
IHE	Indiana Historical Exhibit, Indianapolis
LC	Library of Congress, Washington
MATH	Museum of American Textile History, North Andover, Mass.
MHS	Minnesota Historical Society, St. Paul
MIT	Massachusetts Institute of Technology Museum, Cambridge
NA	National Archives, Washington
NBM	National Building Museum, Washington
PAC	Public Archives of Canada, Ottawa
PICA	Prudential Insurance Company of America, Boston
SMLA	Scott Molloy Labor Archives, Providence, R.I., photos by Miguel Picker
SPNEA	Society for the Preservation of New England Antiquities, Boston
SPNEA, B&A	The Society's B&A Railroad Collection
UBCJA	United Brotherhood of Carpenters and Joiners of America, Washington

For Bobby,
Sonia, Matthew,
and Eva

Contents

Illustrations   ix

Preface   xi

1 With Our Hands: The World of the Carpenter   1

2 Portrait of a Puzzling Industry   16

3 From Artisan to Worker   21

4 The Eight-Hour Strikes: 1886 and 1890   40

5 Union Building   48

6 Inside the Union Hall   67

7 Birth of the Business Agent   78

8 Battling Carpenters: World War I and the 1919 Strike   85

9 Cooperatives: Building Without Bosses   91

10 The American Plan   103

11 Tragic Towns of New England   110

12 Work, Not Relief   122

13 Jobs, Jobs, and More Jobs   130

14 New Tools, New Materials, New Methods   145

15 The Prudential Boom and Beyond   153

16 The Rise of the Open Shop   161

17 An Industry in Transition   188

18 Knocking on the Door: Blacks and Women in Construction   208

19 Who Will Build the Future?   219

Notes   227

Index   237

Illustrations

Through the Eyes of the Howes Brothers   8
Working Lives/Union Life: The Early Years   26
In and Around Boston and Cambridge, 1907–1911   59
Houses, Factories, Ships, and Railroads, 1900–1925   95
Rooftops and Train Tracks: From the Twenties to the Fifties   137
Working Lives/Union Life: Modern Times   174
Working on the T: Photographs by John Laurenson, Jr.   198

I've done everything from woodsheds to fourteen-story highrises.

—Reginald Grover, Local 424

Preface

"It's a tough racket." You hear that phrase a lot on construction sites.

Someone might say it during coffee break, as you try to shake off the freezing wind whipping through a half-completed building on a January morning; or to break the silence after someone falls several stories to serious injury or death; or outside the union hall when you discover a friend hasn't been able to find work in six months. It *is* a tough racket.

But what exactly is it like? Surprisingly little has been written about construction—the nation's largest industry, employing four million people. The popular stereotype of the narrow-minded hardhat, who loves to drink, swear, and fight, is about all we have. Very few people have bothered to penetrate the world of the building trades, to find out what our work is like and what we think about it.

In many respects, carpenters are no different from any other group of workers. We get up in the morning, put in a day's work, come home, and try to handle the daily crises of living. But the world of construction has its own quirks. During building booms, work is plentiful and the money is good. It's hard, it's tiring, but at the end of the day, a carpenter can stand back

and look at what has been built. It's not a world of paper shuffling. The work is tangible and concrete; what's there today wasn't there yesterday. Not every group of workers can drive around their communities, proudly pointing out to family and friends, "I built that."

During slumps, it's a different story. Carpenters wait anxiously with their tools, grasping at anything, even if it's only for a week. Time drags; there are only so many fix-it projects around the house. In the best of times, the work is still insecure—it's in the nature of the industry. Building projects start and finish, lasting from a few weeks to a few years. Carpenters frequently work for two or three employers in a single year. The insecurity has driven many people out of the trade to lower-paying but stable jobs. It's not easy working on a Monday, knowing you might be standing in a line at the unemployment office on Friday.

It's a special world, a peculiar world—one whose rhythms and cycles have rarely been recorded. The physical monuments to the craft and skills are abundant and obvious. Every house, office building, factory, road, and bridge in Massachusetts has heard the ring of the hammer, felt the sharp teeth of the saw, and sensed the familiar hand of the carpenter. But the descriptive or analytic monuments are sorely lacking. "We work with our hands," says John Greenland. "We are far more gifted with our hands than we are with our heads." Greenland's observation unfairly minimizes the poetry and intelligence built into the trade and the people who carry it out. But he is right in the sense that the trade has no tradition of documenting the experiences and perceptions of the men (and now increasingly women) who pass on a language, culture, and set of values from one generation to the next.

The Massachusetts Carpenters History Project was set up in 1983, with a grant from the Massachusetts Foundation for Humanities and Public Policy (a program of the National Endowment for the Humanities), to fill that vacuum and to write part of the unwritten history. The Project has been by, about, and in many ways, for carpenters.

The volume you are holding is not a dispassionate history. How could it be? I am a subject as well as a chronicler of the story. I have made my living as a carpenter for the last sixteen years, twelve of them as a member of Carpenters Local 40. I know the price that has been paid in terms of physical pain and economic insecurity for each wall, each floor, each door, and each piece of hardware that is leaned on, danced on, walked through, and turned. My sympathies are clearly with those workers who are responsible for the society's structures that most people simply take for granted. I have, however, tried to tell the story as fairly as possible, reflecting the variety of events and perspectives that have influenced and altered the working lives of the carpenters of Massachusetts.

The idea for the Project emerged from the centennial of the United Brotherhood of Carpenters and Joiners of America in 1981. The anniversary prompted two full histories of the national union and many shorter local and regional studies of the Brotherhood. This book is, in part, a comparable history of the unions in Massachusetts. Any discussion of carpenters that failed to focus on their labor organizations would be pointless. For over a century, trade competence and possession of a union book have been synonymous. That fusion of craft and union identities has broken down in the last fifteen years as nonunion carpenters have come to outnumber union carpenters, but the role of unionism remains central to the industry. Even so, this book is intended to be more than an account of the ups and downs of the various locals, district councils, and state organizations. I have also tried to focus on the workers themselves, their trade, and their lives in an industry whose structural instability has prompted an elaborate tradition of defensive mechanisms intended to assert some measure of control over a deeply insecure work existence. This book can be read as a history of the carpenters' battle for control at the work site.

This is a conventional history in the sense that it relies heavily on historical research documents, but it also incorporates material from the dozens of interviews of active and retired carpenters that Project members conducted. The decision to work with oral histories was made in order to involve Massachusetts carpenters in both the process and product of the Project. I have tried to weave their observations and perspectives into the narrative so that the history of the Commonwealth's carpenters is told by them as much as possible.

The use of oral histories creates its own set of problems. Memories are often faulty or affected by the changing assumptions of intervening years. Furthermore, each individual carpenter's views are just that—his or her views alone. They are occasionally inaccurate or subject to another interpretation. But as long as that is understood, they are invaluable. For this book is not meant to be a dry citation of numbers and dates. It is a tale of how working lives changed, and how those changes were perceived. It is a history of how carpenters understood and remembered what happened as well as what did in fact happen. The idiosyncratic perspectives of the carpenters we interviewed are as valuable historical nuggets as any primary document that we may have unearthed. I can only hope we did them justice.

There is a plaque in the lobby of the Empire State Building in New York City that lists the names of the craftsmen who built that great symbol of American accomplishments. The decision to commemorate those men was a noble and thoughtful gesture, particularly remarkable because such recognition is so rare. The truth is that the vast majority of building trades workers receive little thanks for their contributions, let alone the majesty of a permanently engraved name. The coincidental fact of being associated with a historic monument should not be the basis for the presence or lack of recognition. After all, the workers who build the most commonplace shopping mall invest the same hours, skills, and care as the Empire State builders. They deserve our thanks as well. It is my hope that this book can serve as a modest plaque to the carpenters who build Massachusetts.

———

As director of the Massachusetts Carpenters History Project, I am ultimately responsible for the contents of this book. That does not mean, however, that other people made no contributions. On the contrary, the Project has been a collaborative effort from the beginning and I owe enormous debts of gratitude.

The Massachusetts Foundation for Humanities and Public Policy funded much of the work of the Project. I am grateful to the Foundation and Kent Jacobson and Faith White, in particular, for their support and encouragement. They attached remarkably few strings to the grant and made clear that they welcomed a popular history and the notion of coupling his-

torical expertise with the articulated experiences and impressions of working people. An important goal of this Project has been to break down the barriers between "observer" and "observed" in the spirit of the British History Workshop as well as the Massachusetts History Workshop and the Brass Valley History Project here in New England. I appreciate the MFHPP's willingness to foster continuing experiments in public history.

As research director, David Goldberg had the difficult but exciting task of developing a research strategy with minimal leads. Since very little has been written about carpenters or construction workers in Massachusetts, Dave was forced to start virtually from scratch. The historical and photographic documentation in this book, particularly in the period from 1880 to 1945, owes much of its existence to Dave's weeks and months in libraries and archives. Dave also conducted many of the oral histories. For someone with no background in the building trades, he proved remarkably adept at grasping the work context of the interviewees. His sympathy and understanding made the interviews enjoyable for everyone involved. I may have written the words of this text, but Dave provided much of the raw material and helped shape many of the ideas. It is no exaggeration to say that the book could not have been written without his participation.

Historians David Montgomery, James Green, and Jeremy Brecher generously devoted many hours of their time from the Project's early stages to the final draft of the manuscript. My sense of labor history has been influenced by their written work, and I have been delighted to discover that they are equally, if not more, provocative and stimulating as coworkers, critics, and friends. Jeff Grabelsky, Maurice Isserman, Mike Kazin, and Barbara Lipski all read and criticized the manuscript. They taught me how generous an act of friendship a careful and exhaustive reading can be. I would also like to thank Dave DuBusc, Mary Eich,

Nordel Gagnon, Mark Hoffman, John Laurenson, Jr., Dick Monks, Miguel Picker, Ellie Reichlin, Rob Snyder, and the organization City Life/Vida Urbana.

The Project was an independent effort, but owes much to the support and good will of officers and members of the United Brotherhood of Carpenters and Joiners of America. Andy Silins and Bob Bryant provided important assistance all the way through. As president of the Massachusetts State Council of Carpenters, Bob contacted all the local unions in the state and encouraged them to develop lists of retired carpenters for our interviews. Andy, the general agent of the Boston District Council of Carpenters, has been a reliable source of encouragement over the years. Both of them made it clear that they endorsed the idea of a thorough and responsible history of the union. Joe Power of Local 40 jumped into the Project with both feet. He conducted numerous interviews, commented on early chapters, and always helps me think about the world of work we share.

In addition, I would like to thank Massachusetts Carpenters Union officers Jim Martin, Mike Molinari, and Donald Shea. Help came from the International Office of the Brotherhood in Washington, D.C., through John Rogers, Roger Sheldon, Beverly Breton, Linda Coller, and Theresa Threlfall. Many, many more people in the union have known about and supported the Project in a variety of significant ways. Still, all the opinions expressed in this book are mine alone.

Last, but most definitely not least, I want to thank all those carpenters who allowed us into their homes, their lives, and their pasts. It was a pleasure on both sides of the tape recorder. Carpenters are not always accustomed to having their views being taken seriously enough

for inclusion in a book. As a result, the people we interviewed eagerly opened their doors and shared their memories. For those of us in the Project, it was equally enjoyable. We got a rare and gratifying chance to listen to a generation's experience unfold in front of us.

Throughout this book, quotations without source notes are from the collection of oral history interviews conducted by the Massachusetts Carpenters History Workshop. The names of all the interviewees are listed below. All quotes from documentary sources follow a standard citation format.

Carpenters interviewed were (in alphabetical order): Arthur Anctil, Nazadeen Arkil, James Audley, Carl Bathelt, Cliff Bennett, Leo Bernique, Ellis Blomquist, Gordon Boraks, Angelo Bruno, Bob Bryant, Faith Calhoun, Joe Corbett, Leo Coulombe, Richard Croteau, Angelo DeCarlo, Joseph Emanuello, Fred Ernest, Edward Gallagher, Wilfred Goneau, L. P. Goodspeed, John Greenland, Reginald Grover, Tom Harrington, Ed Henley, Harold Humphrey, Sharon Jones, Bob Jubenville, Ernest Landry, Joseph Leitao, Bob Marshall, John MacKinnon, Mitchel Mroz, Enock Peterson, Joseph Petitpas, Thomas Phalen, Joe Power, Oscar Pratt, Harold Rickard, Tom Rickard, Chester Sewell, John Short, Bob Thomas, Al Valli, Barney Walsh, Bob Weatherbee, Paul Weiner, Michael Weinstein, and Mary Ann Williams.

We also interviewed several people with other connections to the industry. They are: Omar Cannon, Felix J. Conte, Leo Fletcher, Wilbur Hoxie, Eric Nicmanis, Stephen Tocco, Chuck Turner, and Manny Weiner.

There were many more people who wanted to be interviewed than we were able to connect with. I apologize to them, knowing that the stories we missed were really our loss.

As I pass up and down [my city's] streets I see in many places the work my own hands have wrought on her buildings and I feel that in a sense I am a part of our city. My strength and whatever skill I possess are woven into her material fabric that will remain when I am gone, for Labor is Life taking a permanent form.

—Walter Stevenson, union carpenter, 1930

If you try to kneel down in the Riverside Church in New York, you break your nose on the pew in front ... Who kneels down in that church? I'll tell you who kneels. The man kneels who's settin' the toilets in the restrooms. He's got to kneel, that's part of his work. The man who nails the pews on the floor, he had to kneel down. ... Any work, you kneel down—it's a kind of worship. It's part of the holiness of things.

—Nick Lindsay, carpenter
in Studs Terkel's *Working*

A couple of times a year we always had a get-together. It didn't cost the union much. Sometimes we'd have a band. Your wife got to meet my wife. We knew we weren't all a bunch of tramps because we all got dressed up. We didn't always go around with mud in our shoes and dirt on our clothes.

—Angelo Bruno, Local 402

1

With Our Hands: The World of the Carpenter

"I don't care if I am doing the best finish work or making a concrete form," says Joseph Petitpas. "I take pride in what I am doing." Bob Thomas agrees: "You learn every day, you learn every minute." In today's mechanized, routinized, and often alienating blue- and white-collar work places, the carpentry trade can be a refreshing alternative. Though the craft may not be as demanding as it once was, carpenters still have to use their wits and solve problems on a daily basis. "Finish work is a challenge," claims Leo Bernique. "It takes patience, skill, and know-how. There's a certain time to be rough and a certain time to be very gentle. I wouldn't have traded jobs with anyone in the world while I was doing that kind of work."

For many, the work never became monotonous. "You work eight hours and you love it," explains Ernest Landry. There is satisfaction knowing that the product of a carpenter's sweat and labor stands long after the job is done. Landry likes to take his children to the Prudential Center to show them where he worked on each of the fifty-two floors, just as John MacKinnon enjoys looking at cabinets he installed in department stores. Tom Rickard appreciates the teamwork it takes to complete a building—"it makes you proud to know that you were a part of it." Tom Harrington identified with his projects so completely that he nicknamed each of his sixteen kids after the job he was on at the time of their births. That way, he jokes, he knew when "Portsmouth Memorial" was squawking.

Most carpenters were raised with tools in their hands; few came to the craft cold. Harrington's experience is typical. "We would drop by jobs when I was a kid, watching my father. We used to take his lunch down to him. So we got to know the fellows, the work, the trade right there. We knew the expressions, which is sometimes a language in itself." When he was eight years old, Leo Coulombe went out shingling with his father. "Dad showed us from *A* to *Z*. From the foundation to the painting to the wallpapering. He taught us the whole thing." Ellis Blomquist started working summers and vacations for his father at the age of twelve. Sometimes the line between work and home was fuzzy in these "family apprenticeships." Every Sunday, says Harrington, his father would lay out his blueprints in the parlor and explain them to the children.

Joel Leighton of the Massachusetts Associated General Contractors (AGC), an organization of building employers, once commented that only one-quarter of the journeymen carpenters employed by AGC firms went through a formal union apprenticeship. Other observers confirm that estimate.[1] There have always been alternate, or "backdoor," routes into the craft. For many years, a private organization called the Smith Charities sponsored training programs for young men and women in the Pioneer Valley. Boys interested in becoming carpenters were indentured to an employer and given $300 by the Charities if they finished their training. More often than not, carpenters skipped structured programs altogether. They picked up the trade wherever they could and joined the union as journeymen. Ernest Landry followed this path, passing a test administered by several union members in 1947.

Locals had standing committees to check the craft knowledge of the applicant. "You had to know quite a bit," remembers Joseph Emanuello. He joined the union in 1946, but not before demonstrating basic knowledge of the trade. "You had to know how to frame. That was the big thing. Could you frame a house? Could you frame stairs? Could you cut stairs? Could you cut rafters?" James Audley had been in a trade union in Galway, Ireland. When he came to the United States, he took a test to enter the Carpenters Union.

There was seven of us young fellows that night. They took each of us into a separate room and asked us questions. They said, "What figures would you take on your framing square to make a wood miter box?" They asked me about the standard height for sawhorses, and how to prove a foundation layout was square. They asked me how to scribe panelling when corners aren't plumb. Anyhow, I answered all the questions and got my card that night.

Along with whatever admission procedures locals insisted on, new members inevitably faced informal on-the-job tests. Bob Bryant worked for his father, a small contractor, from the age of fourteen. When he joined the union as a journeyman, he still had to prove his abilities to the foremen and his coworkers, always skeptical of a newcomer. "They could pretty much tell if you knew what you were doing if you could build a sawhorse and how long it took you, if you had a pair of overalls, a full toolbox, and if you carried it down to where you were working. If you didn't have a full toolbox within walking distance, you'd know about it pretty quick."

Ultimately, learning the job came down to, in Richard Croteau's words, "stealing the trade." Harold Rickard confirms the necessity

Leo Bernique
By Nordel Gagnon

Mitchel Mroz

for independence. "I just picked it up on my own." Since most carpenters grew up in mechanically oriented homes, they had some idea of what to look for and whom to rely on. "You'd see what was going on around you and you'd talk with good journeymen," Leo Bernique says.

"There was no teacher," agrees Al Valli. "You had to learn yourself." But what made the difference between a productive or a frustrating apprenticeship was the presence or absence of informal teaching. For all his claims of self-reliance, Valli admits that he learned quickly because a job superintendent took him under his wing. Many carpenters have fond memories of the veterans who shared their craft wisdom. "The old-timers were helpful," Bob Thomas remembers gratefully. "They'd try to teach you things that they knew, so long as you weren't a smart aleck." Thomas Phalen started out with W. J. Hanley of Fitchburg in 1927 and worked steadily with the company until the winter of 1932–33. "I owe a hell of a lot to some of those old fellows, French-Canadians. They couldn't read or write, but they could read blueprints. They were wonderful mechanics and they taught me a lot."

For every pleasant encounter, there were the

tougher initiations as well. Apprentices, as the saying goes, were to be seen, not heard. Older carpenters often enjoyed the ritual of breaking in the newcomer to see if he could "take it." Mitchel Mroz describes his first job in 1939.

I was half laboring and half doing carpentry. They'd send me for stuff and I'd go get it and bring it back and wait for something else. They'd see me standing there and send me for something else they didn't need. So the boss came by and asked me, "What the hell you bringing that for?" I said, "They wanted it." He said, "Take it back. I know what they're doing. You're watching them. They're scared you'll take their jobs away since you're learning the trade. They don't want you to do that." That's the way it was. They wouldn't even try to help you. They'd give you the biz.

Apprentices have always done the work no one else wanted—the heavy lifting and the repetitive tasks. They paid their dues to join the union and did the same on the job. In a history of New York carpenters, an upstate house builder talked about some of the apprentices' duties in the 1920s:

When the lumber was brought to the job by horse and wagon and got stuck in the mud, they would dump the load right there and say, "Contract reads—Delivery as far as good roads go." Then the apprentice had to hump the whole damn load up through the mud to the job site, sometimes 200 or 300 yards away.[2]

When Bob Weatherbee came into the union in the early 1960s, matters had not improved. "You were treated like dirt. Whatever they could do to put you down, they did. They made a point out of making sure that you knew you were an apprentice."

Today as in the past, the fortunate novices "partner up" with veteran carpenters who teach them the tricks of the trade. The less fortunate spend four years stuffing walls with insulation, lugging supplies to and fro, and taking the coffee orders. The particular destiny is largely a matter of chance, depending on the needs of the job or the whim of the foreman. Observation and imitation remain the true pillars of a carpenter's apprenticeship. As Leo Bernique said of his introduction to the trade, "You learned by keeping your eyes open."

Older carpenters teach their younger co-workers how to be the right "kind" of carpenter as well as how to be a competent craftsman. The passing on of the craft culture and identity is as crucial as the correct grip on a handsaw. For without craft pride, a lifetime in construction can be a difficult and stressful experience. It is a life filled with constant layoffs, injuries, insensitive employers, and chronic economic insecurity, the veterans warn the apprentice, and if that message is ignored, the frigid winters of Massachusetts drive the lesson home.

In 1879, a carpenter called for higher wages because "our New England climate requires a greater variety of clothing than any other part of the country."[3] The weather has not improved in a hundred years. "You've got the wind whistling around your ears and you're shivering and shaking," remembers Thomas. "Many times you feel like saying the blazes with this and walking away." The thermometer may jump or slide, but the outside work—framing, concrete forming, roofing—goes on. Harrington describes pile driving: "They want those piles driven whether there's snow or rain or sun. They say the machine is driving the pile in there and the machine will keep going regardless of the weather."

"There is no such thing as steady employment in the building trades," commented Arthur Huddell in 1921. "There never was and never will be until the skies stop leading snow and rain."[4] The weather, the seasons, and the boom-and-bust nature of the building cycle all conspire to force some hard choices. "When you're loafing," Rickard says, "you try to scrape up what you can." Every carpenter has loafed at one time or another, even during the boom years when the problems of seasonal employment are moderated. In 1982, a strong year for building in Massachusetts, the state's construction workforce worked a total of 24 percent fewer hours from January to March than they did from July to September. The average number of building trades workers that year was 78,879, but the difference between the number employed in March and August was almost 18,000.[5]

Finding a job in a lean period can take a matter of years; sometimes it just means holding out until the spring building season. "Years back," says Coulombe, "you didn't find any work after Thanksgiving." Richard Croteau used to make sure his bills were paid up by November or December. He cut ice or fixed looms in the Lawrence mills in order to earn extra cash. The only thing that kept his family going in the winter, he claims, was credit from the corner grocer and the coal yard. "It was toughest during the wintertime," recalls Harrington, who was born in March. "A lot of times my birthday presents had to be picked up sometime in June. My family didn't forget about me; we just accepted the fact."

If the pickings are too slim, carpenters leave their families and travel or "tramp" in search of work. Many people drift in and out of the industry altogether. A Department of Labor study of 79,000 construction workers showed that only half were able to support themselves on construction earnings alone. The rest relied on other sources of income—as factory hands, salesmen, or service workers.[6]

Craft pride is stretched thin by corner-cutting contractors eager to save a dollar. European craftsmen are often surprised by the atmosphere on construction sites in the United States. James Audley arrived in this country in 1948 and was immediately struck by the contrast. "In Ireland you couldn't get any of your work fine enough. Here I noticed as soon as I

Richard Croteau
By Nordel Gagnon

came that it was just mass production, just speed." Angelo Bruno worked for a ceiling contractor who offered an incentive plan to company foremen—if a job came in under the bid, the foreman got a percentage of the extra profit. He describes the consequences of policies that reward greed rather than quality: "We had a school in Connecticut that was a tongue-and-groove staple job. They were putting two staples in the tile, instead of six. That's what the foreman told the guys he wanted. Before the job was even done, the weight and vibrations from the upper floors worked it loose and it all came down."

As overall building costs have mounted, the pressure for speed has intensified. Petitpas thinks "the pace is a lot quicker, they want a lot more now. Then they just wanted you to keep busy. Today they want you to keep busy plus." Mitchel Mroz points to the late 1950s and early 1960s as a turning point. "Now it's the heck with it, it's good enough, get it up." Older carpenters mourn the passing of an era when, as they see it, quality mattered, and marvel at the logic of "fast and dirty" construction methods that require constant repair. "How come they have time enough to do it a second

time," wonders Bruno, "but they don't have time enough to do it right the first time?" Greater speed not only sacrifices quality, it also increases danger. "They didn't want you to have guardrails," says Rickard bitterly. "They wanted you to use a lousy two-by-four to walk on." Some companies, he goes on, "didn't care. They didn't give a hoot if you got killed or what. All they wanted was it done."

Construction can be a deadly force. The industry only employs 5 percent of the total workforce, but is nonetheless responsible for 20 percent of all job-related fatalities. In addition, construction workers have the highest rate of injuries and illnesses of any group of workers. The missing finger, the chronically aching back, the broken bones are all sure signs of the carpenter's trade. Petitpas lost the tip of his finger handling a piece of plywood on a roof on a windy day. A concrete overhang broke under Harrington's feet, fracturing his neck and back and forcing him into a twenty-five-month hospitalization. At one time or another, nearly every carpenter has taken a fall or had a brush with serious injury. Harold Humphrey vividly recalls the sparks flying around him when he stood on a steel form that touched a live 440 volt wire. "I should have been cooked right there," he now laughs. Like most carpenters, Bob Weatherbee has seen fatal accidents on the job. "On one job a cement finisher was finishing off a building where he had a foot to stand on. He stood back and over the building he went, about ten stories."

Carelessness causes some accidents; the "tough" construction worker too often looks the other way when faced with unsafe practices. Others are an inevitable part of a risky business. But the majority of accidents are entirely unnecessary. Petitpas angrily recalls years of fighting to get the most basic safety equipment. "To get the company to put up guardrails and ladders, that's where the headache is. You can demand it and they just take their time. When somebody gets hurt they

Joseph Petitpas
By Nordel Gagnon

hurry up and put up a ladder so there will be one there when OSHA shows up." Serious accidents momentarily stun the work crews. Talk drifts uneasily to similar incidents in the past and the likelihood of more in the future. As frightening as they are, the accidents can be accepted—it is the cavalier employer attitudes that burn. Petitpas tells an all-too-common story of the superintendent who complained when a carpenter left the job in order to take another severely injured worker to the hospital. And the worst fear—a fatal accident—is never far away. "I worked a job in the 1950s where two carpenters were killed," relates Bob Routen. "They were looking out from the side of a building when a cable snapped. It took off both their heads. The *Globe* had two little paragraphs on it. I quit the next day."

Carpenters just keep going, muttering under their breath or complaining mildly to each other. They put on another layer of clothing, wait for a job to turn up, or shrug their shoulders and continue working. For those who spend a lifetime in the trade, the benefits—the sense of mastery over the tools, the pride in a job well done—can outweigh the dangers and insecurities. The occasional chance to build

something right, to figure out the best approach and carry it out under safe and pleasant working conditions, can make up for weeks and months of slogging in the mud, performing mindless and repetitive work, or fighting a "pusher" foreman.

The insult to self-respect, to craft pride, cuts deeply. Whether the cause be a new labor-saving device, prefabricated material, or a contractor willing to trade speed for quality, every carpenter bemoans the loss of trade skill and is convinced that his generation is the last one to "really" know how to build. In 1829, an artisan grumbled that the trade had been divided into half a dozen subtrades and that "every working tool is simplified."[7] Seventy years later, a Worcester carpenter attacked the rise of "wood butchers, men who come from the backwoods with a kit of tools comprising a saw and a hatchet."[8]

The story has not changed. Chester Sewell believes the difference between younger and older carpenters today is that "less skill is required now." "When I was an apprentice," Tom Phalen points out, "you were taught to do everything. Today you don't put a window frame or door frame together. It comes all ready and you just stick it up." Mroz echoes this view. "You can't get these younger carpenters to cut rafters, lay out stairwells. They go on a job and they're told what to do and they pick up what they can."

Mroz is partly right. In training programs today, young apprentices frequently question the point of learning how to cut roof rafters or stair stringers. Why bother, they ask, when factory-built roof trusses and stair units have made those skills obsolete? A novice carpenter can spend years in the trade building forms for concrete, framing walls with metal studs, and installing drywall, without ever getting a chance to work with a piece of finished wood. Still, relative to other traditional crafts, building trades workers have retained significant skills. Their industry evolved in fits and spurts but never succumbed to the assembly line. A nineteenth-century carpenter, walking onto a construction site today, might be bewildered by the heavy equipment, power tools, and modern building materials, but he would not be entirely lost. The hammer, the handsaw, the level, and the square—time-tested emblems of an ancient craft—are still indispensable.

Will they continue to be? The construction industry today is in the throes of a dramatic and potentially far-reaching transformation. As a result of a major reshuffling of homebuilding techniques, low- and moderately priced single-family houses are now almost exclusively constructed in factory settings, leaving only the expensive custom house in the hands of con-

ventional on-site builders. Multistory residential, commercial, and heavy construction remain insufficiently standardized for the limited options of the factory system, but more and more individual components—cabinets, door and window assemblies, even finished walls—are preassembled behind plant gates. The carpenter's work is changing; some new skills are required, but overall the craft will become less complex and more specialized. As this history suggests, such periods of industrial reorganization in construction have happened before. A constant theme of this history is how and to what degree carpenters have been able to adapt to the shifting sands of their work environment.

Through the Eyes of the
Howes Brothers

In 1888, Alvah Howes opened a photography studio in Turners Falls, a small town nestled in the Berkshire hills of western Massachusetts. For the next twenty years, Howes and his brothers Walter and George eked out a meager living in the "view business," as Walter called it.

The Howes brothers were neither self-proclaimed social chroniclers nor courtiers to the wealthy—their aims were more modest. Constrained by the realities of small-town life in the Connecticut Valley, they simply took pictures of anyone who would pay their fees. In 1889, the *Turners Falls Reporter* ran an announce-

ment for the Howes studio under the headline: HAVE MAMA'S PICTURE TAKEN. The same advertising copy reappeared every week with just one change in the banner—LITTLE SISTER, GRANDMA, PAPA, BABY—until every conceivable family connection had been exhausted. With no more human subjects available, the ads turned to KITTIE and THE DOG for suggested portrait sittings.

Turners Falls and its neighboring towns were too small to support the three photographers and their families. Throughout their careers, the Howes brothers—particularly Walter and George—

8

roamed far afield in search of clients. The precarious nature of their business drove the brothers from town to town, hoping to find new faces willing to pay for a moment in time frozen by their cameras. The life of the itinerant photographer may have strained the resources of these men, but it has left us a remarkable collection of photographs—a visual social history of turn-of-the-century western Massachusetts.

The subjects are astonishing in their diversity—rich and poor, farmer and factory hand, men and women, in group shots and individual portraits. The Howes had an unerring feel for context, par-ticularly in their depiction of peo-ple and their work. The tobacco hand is shown in the fields, the logger in the forest, the laundress in her factory, the painter with brush in hand, and, above all for our purposes, the carpenter in his world of work.

Fortunately, the bulk of the Howes brothers' glass-plate nega-tives remain intact, protected over the years by the dry air in the attic of the Zachariah Field Tavern in their hometown of Ashfield, Mas-sachusetts. In the early 1960s, the collection was donated to the Ashfield Historical Society. After years of painstaking preservation and cataloging, over 21,000 Howes brothers negatives are now ac-cessible to students of late-nineteenth-century New England. Some of their photographs can be seen in *New England Reflections, 1882–1907: Photographs by the Howes Brothers*, a 1981 book edited by Alan B. Newman and published by Pantheon Books, New York. The pictures on these pages are published here for the first time, with the permission of the Ashfield Historical Society. © 1979, 1981 ASHFIELD HISTORI-CAL SOCIETY

9

11

12

The building industry, disjointed, disorganized, with a clientele suspicious and largely uninformed of its complexity, with an architectural profession almost equally uninformed and clamoring for a recognition of superior knowledge of the problem which it never possessed and cannot maintain, with the banking and lending institutions throughout the country taking no stand for a stabilized industry, but relying on an assumed satisfaction with plans and specifications made in a medium they do not comprehend and written in a technical language that they cannot fully understand, with bonding companies as ready to insure the performance of an inexperienced beginner as an experienced builder, so long as the premium is paid, with importunate novices clamoring that they can build cheaper than anyone else, with the sheriff waiting in the treasurer's office while frantic collections are being garnered in the banker's office to stave off for another brief period the hand of bankruptcy that overtakes fifty percent of his kind in every five-year cycle—is it any wonder that we have never met seriously to stabilize labor relationships? Is it not a great wonder that absolute anarchy in the industry does not completely overwhelm it?

—William Starrett, 1928[1]

2

Portrait of a Puzzling Industry

In all fairness, "absolute anarchy" was too strong a phrase, even for an exasperated building contractor. William Starrett's portrait of his environment may have been overdrawn, but he did, nonetheless, capture important elements of a business filled with a bewildering array of actors—all with pivotal roles. The very number of players—architect, banker, contractor, developer, engineer, right down the alphabet to worker—combined to create an industry without a clearly defined center. Starrett, a member of a prominent New York building family, was neither the first nor last to harp on the vacuum at the core of the industry. This universe of interdependent constellations perpetually irked business analysts more comfortable with the organizing logic of a dominating institution or figure.

In a 1947 cover article entitled "The Industry

Capitalism Forgot," the editors of *Fortune* lashed out at the "feudal character" and "picayune scale" of construction's homebuilding wing. It was, they sputtered, "the one great sector of modern society that has remained largely unaffected by the industrial revolution."[2] Much of their irritation focused on the limited ability of building contractors to chart their own destinies as compared to employers in other industries. Local or regional in scope, pre–World War II American construction employers shared little of the authority or august aura of their peers in basic manufacturing. There were no Carnegies, Rockefellers, or Fords erecting offices, bridges, or houses. Instead, even major builders like Starrett portrayed themselves as handcuffed by their occupational environment.

Starrett and *Fortune*'s editors believed a

16

qualitative leap in employer size was the appropriate prescription for construction's supposed ailments. And, in fact, their plea for bigness has, to some degree, been answered in the past forty years. Construction giants, such as Fluor, Bechtel, and Brown & Root, currently operate on a global scale; even homebuilding now has billion-dollar firms. The 1982 Census of Construction Industries, most recent of the Commerce Department's regular five-year surveys, reported that 71 percent of the nation's total building business receipts belonged to just 4 percent of the construction firms. In Massachusetts, the figures are even more extreme. Three percent of the state's construction companies performed 91 percent of the work.[3]

These data can be misleading. The top 3 or 4 percent do not represent a handful of massive corporations exerting a stranglehold on the market. Unlike most other industries, construction remains a relatively easy field to enter. With a grand total of 1.4 million construction establishments in the United States in 1982, 4 percent adds up to 52,668 companies—hardly a monopoly. Similarly, the top 3 percent in Massachusetts includes over a thousand firms. Certainly, the industry has become increasingly centralized, but the figures also reveal the remarkable persistence of the small "mom-and-pop" outfit. The census indicated that 932,608 companies across the country functioned without *any* employees in 1982 and collectively managed to carry out $41 billion worth of business, 11 percent of the nation's total construction volume. These individual builders or working partners offer no serious threat to the multi-billion-dollar giants, but their existence is testimony to the continuing niche for small builders.[4]

More important for the purposes of this book, construction workers *experience* their work environment as decentralized. Giant payrolls are the exception, rather than the rule, even among the larger firms. Most building employers still operate in local or regional markets. Sixty percent of the construction workforce (craft workers as well as office staff) in the United States work in companies with fewer than fifty employees. Even more astonishing, fully one in four is employed by a contractor with fewer than ten workers.[5] The combination of working within small or medium-scale operations and the accessibility of self-employment has had an enormous impact on the historical consciousness of American construction workers. Over the course of their working lives, a significant number have been, in turn, employed by others, self-employed, or (less likely) employers.

The vast majority of building trades workers collect a paycheck. But within that framework, tradesmen and -women who will never open their own businesses have temporary escapes from the status of a hired hand. Frequently, they will work on their own in the evenings and on weekends. At some time, most carpenters will offer bids on jobs, even if it is just a matter of repairing a porch, installing kitchen cabinets, building a deck, or adding a room. They will order their own building materials, pay a neighborhood teenager, and fight to collect an overdue bill. As a result, they will have faced, from a worm's-eye view, some of the issues confronting their employers. Whether it be working "side jobs" or buying a pick-up truck for tools and materials, from time to time most construction workers will grapple with a fuzzy self-definition that straddles strict class boundaries. Autoworkers, check-out cashiers, and nurse's aides are unlikely to open a factory, supermarket, or hospital, but carpenters and other building trades workers can harbor small-scale designs of independence within their occupation. The chances of truly "making it" may be remote, but the nine hundred thousand self-employed builders counted by the census do underpin a construction culture in which distinctions between employer and employed are sometimes blurred and the common concerns of everyone who builds for a living can be exaggerated.

Despite its internal disorder, decentralization, and vestiges of occupational mobility, construction has a long history of organization—among both workers and employers. In many ways, trade unionism blossomed among American building trades workers because of, not in spite of, the decentralized character of the industry. The indispensable level of the workers' skills gave them extensive bargaining leverage against employers with limited financial resources and uncertain political cohesion. Individual builders rarely had the luxury to withstand lengthy strikes by a workforce that was not easily replaced. Thus, both sides had a compelling incentive to organize their constituencies thoroughly.

The nature of the unionism that evolved in the building trades also flowed from the makeup of construction. Craft unionism—the organization of workers on the basis of a common trade skill—spoke to the carpenter's need for a secure identity in an insecure climate. Craft unionism offered a haven, creating a semblance of order out of chaos. Membership in a building trades union represented more than an automatic economic move. It functioned as a validation of an occupational choice, providing extensive social and cultural services, constantly reaffirming the value of a life spent handling wood. From the beginning, pride in one's craft and one's labor organization went hand in hand not only for the union organizer but for the typical working carpenter.

The form of craft unionism symbolized the kind of solidarity that existed for carpenters—one that was based on a feeling of kinship with fellow carpenters rather than the generalized working population. The union's hand of friendship was usually extended to other construction unions but less consistently toward labor organizations outside the industry. At its most extreme, this standoffish stance became outright hostility toward unskilled workers. During much of the highly charged debate on craft versus industrial unionism in the first three decades of the twentieth century, the national leadership of the United Brotherhood of Carpenters and Joiners of America argued fiercely against the wisdom of organizing on an industrial basis.

In order to succeed, craft unionism in construction required firm and total control over the labor market. A local union's bargaining strength was directly related to the complete organization of the carpenters in its community. Unorganized skilled carpenters represented a serious threat to the union, an opportunity for contractors to staff their sites adequately with nonunion tradesmen. The need for control over the labor supply also drove the unions to adopt diametrically opposed strategies. Once a local signed up the bulk of its potential membership, the open-arms approach gradually gave way to policies based on limiting the labor supply and excluding further competition for scarce jobs.

Union attitudes toward apprenticeship have always demonstrated the perceived need to control access to the construction site. Naturally, union members sought to limit rapid influxes of carpenters from any source, but since the apprenticeship system was the only formal means of entry into the trade, it provided a convenient handle for regulations. The threat of an unrestricted number of openings has frightened older carpenters for more than a century. Young, strong, agile, eager, and low-paid apprentices present an obvious and attractive alternative to higher-paid journeymen past their prime for cost-conscious builders. Construction unions quickly moved to establish fixed journeymen–apprentice ratios.

As early as 1868, the president of the National Union of Bricklayers announced: "It is only by limitation of apprentices that we can hope to regulate the demand for our labor."[6] Such statements invariably set off howls of indignation from employers and free labor market defenders. Union leaders were accused of seeking labor monopolies and artificially raising prices through the exclusion of unskilled and semi-skilled workers. Unionists responded, defensively at times, that no skilled mechanic would ever remain in a trade that was perpetually seeking to replace him.

In a 1903 editorial in the *Carpenter,* the journal of the United Brotherhood, General Secretary Frank Duffy suggested that the number of apprentices should be tied to the amount of unskilled work built into the craft.

The limitation of apprentices should be directed, as far as possible, toward eliminating the class of men who to-day merely exploit the trade, lower the standard of workmanship, discredit the just claims of good mechanics and demoralize the entire craft. . . . In limiting apprenticeship in this manner, the unions would only be following the example of many trade associations and professions that now legally prescribe the methods of entering their ranks and enforce compliance with their rules.[7]

The economic need to exclude excess carpenters dovetailed neatly with the cultural need to create a tightly knit community of carpenters with a clear collective identity. In Massachusetts as in other states, this process of craft union/craft identity building was hastened by the minimal ethnic, racial, and sexual divisions within the workforce. Carpentry always has been and still is a predominantly white male trade in which craft knowledge is passed from father to son. Black carpenters were scarce in Massachusetts until fairly recently; there were no Rosie the Carpenters in World War II; and even ethnic differences among the white men faded in the face of the largely native-born, Canadian, or northern European origins of the workforce. While this homogeneity has been one source of tradesmen's mistrust of affirmative action in the industry and "outsiders" in general, it has also played an important role in cementing bonds of solidarity within the trade.

Thus, the very structure and dynamics of the construction industry have bred seemingly contradictory responses from the people who labor in it. Fierce individualism and a strong collective consciousness are married in the brand of craft unionism peculiar to the building trades. Workers view their employers as opponents within the bargaining context, as allies in relation to the larger social and political environment, and as role models for those seeking a route out of the drudgery of wage labor. The unions combine a warm, supportive, even nurturing, atmosphere for those "inside the family" with a studied indifference toward those outside the industry.

What has held all the disparate strands of the carpenter's life from spinning off into a formless existence is the glue of the self-conception of a carpenter as a culturally valuable entity. Craft pride has been at the core, expressed in a variety of ways—in the turn-of-the-century carpenters with white coveralls over their white shirts and ties, in the ample toolboxes overflowing with finely honed saws and chisels and esoteric tools that are included "just in case," and in the ultimate compliment of being considered a "good mechanic" by one's peers. Respect must be earned, but it is ultimately guaranteed for those who choose to stick it out. The inclusive nature of the craft union culture has allowed the continuation of one of the most astounding characteristics of the industry—the uniform wage for *all* journeymen in the union setting. In a society that reveres pay-scale hierarchies and provides status rewards on the basis of income differentials, union carpenters and other organized building trades workers have offered a distinctive and almost subversive model for compensation systems.

The American carpenter today performs many of the same tasks as his predecessors. Every morning he shows up at the construction site ready to erect the foundation, structural frame, partition walls, or interior finish. But developments in the industry are challenging the relevance of the once remarkably constant self-definition. Technological innovations have simplified the work, rendering the carpenter more of an installer than a fabricator and planting doubts as to whether the future holds room for an entire occupational category of "good mechanics." The rise of the open shop in construction and the overall antiunion political climate diminishes the once undisputed position of the craft union as the organized expression of the carpenter's needs.

This book recounts the history of carpenters in Massachusetts. The contemporary image of crisis that pervades the field is not without precedent. On the contrary, with the exception of a few stable decades in the post–World War II era, carpenters, their trade, and their labor organizations have been under perpetual attack. Craft pride has served the carpenter well, as a bulwark against the slipshod contractor, the reckless foreman, labor-saving devices, and the fundamental insecurity that has driven so many out of the industry. That sense of pride, however, has also been in a chronic state of tension in a world of work that simultaneously reinforces and undermines the very craft that carpenters depend on. New conditions, new circumstances require new and creative approaches, but a look to the past also reminds us that few of the battles the modern carpenter faces have never been taken up before.

It is not a long period since some of our opposers made it a rule to furnish a half pint of ardent spirits to each man, every day, for no other purpose than to urge the physical powers to excessive exertion. . . . Now we are told that excessive labor is the only security against intemperance . . . we have only to say, they employ us about eight months in the year during the longest and the hottest days, and in short days, hundreds of us remain idle for want of work. . . . When the long days again appear, our guardians set us to work as they say, "to keep us from getting drunk." No fear has ever been expressed by these benevolent employers respecting our morals while we are idle in short days, through their avarice.

—Boston House Carpenters, Masons, and Stone Cutters circular, 1835[1]

3

From Artisan to Worker

In 1633, the court of the Massachusetts Bay Colony set a ceiling of two shillings a day on carpenters' wages. According to John Winthrop, first governor of the colony, the court was forced to take this action because "the scarcity of workmen had caused them to raise their wages to an excessive rate, so as a carpenter would have three shillings the day." The Puritan Winthrop worried that too much income would lead to idleness and bad habits. "They spent much in tobacco, strong waters, etc.," he wrote in his diary, "because they could get as much in four days as would keep them a week."[2]

The colonial craftsman was well paid. Even with the court's limit, a carpenter in Massachusetts regularly earned more than twice as much as his English counterpart.[3] His relative affluence stemmed from a high demand for new houses, commercial buildings, wooden ships and wharves, and a chronic shortage of skilled labor. The carpenter who brought his trade and tools from England was a man with considerable bargaining power.

The building industry in the budding towns of New England was modeled on the European guild system. Borrowing designs and building

methods from the old country, master carpenters opened up shops and sold directly to consumers. As their reputations and orders expanded, masters hired journeymen and apprentices to help carry out the work. Masters, journeymen, and apprentices alike handled the tools. The division between the groups was not one of permanent status, but of age, years of experience, and level of skill. Barring unforeseen circumstances, an apprentice who stayed with the trade could expect, over time, to become a master. The rules were unwritten, but tradition held that masters looked out for the long-term welfare of journeymen and apprentices and passed on the "art" and "mystery" of the craft to the next generation.

The prescribed route of entry to the craft was through apprenticeship, a system that originated in the guild traditions of Europe and has been used with varying degrees of success ever since. Apprentices in colonial America were "indentured," a legal process by which the guardianship of young men was passed from parents to master craftsmen. In October of 1640, for example, Thomas Millard was indentured for an eight-year apprenticeship to William Pynchon, a Springfield carpenter. In ex-

21

change for his work, Millard received room, board, and clothing and the promise of "one new sute of apparell & forty shillings" upon completion of his training. Millard left four months before his term was up. Though he lost his promised forty shillings, he did get his suit.[4]

For all the bustle and trade of the urban centers, colonial Massachusetts remained an agricultural society. As late as 1820, only one of every fourteen Americans lived in cities with a population over 2,500. The farm was the basic economic unit and most families either were self-sufficient or bartered with their neighbors. Farmers who knew how to build often headed for town after the fall harvest and stayed until the spring planting. The threat of the farmer/carpenter's competition was a sore point for urban carpenters. "A Farmer ought to employ himself in his proper occupation," huffed a New York mechanic in 1757, "without meddling with Smiths, Masons, Carpenters, Coopers."[5]

Conversely, if a new house or barn needed skills beyond the farmer's or the community's talents, there were builders who roamed the countryside offering their services. These all-purpose carpenters who designed and built houses by pattern books, local traditions, or the seats of their pants represented an American alternative to the European guild tradition. Typically, a farm family would contract with one of these men to put up a structure in exchange for room and board and a small sum at the end of the job. In addition, the farmer supplied the materials and provided whatever unskilled help the carpenter needed.

Just as the traveling carpenter had to improvise with the materials and terrain available on each project, he was also forced to develop other sources of revenue after each building season wound down. Unlike the urban master mechanic who retired to shop work in the colder months, the rural builder needed a wider variety of skills to survive. For example, in the early 1800s many carpenters found part-time employment in the emerging shoe industry in Lynn, laying down their hammers and saws to pick up the tools of the shoemaker every winter.

In 1774, carpenters in Boston issued a "book of prices," a slim volume of rules of work and a list of recommended prices for various types of work. This book (and similar manuals in other cities) was intended to stabilize building practices by cutting down on competition between carpenters. While publication of piece rates may have bred cooperation among masters, it unintentionally promoted conflict in the presumably harmonious relationship between masters and their journeymen and apprentices. "The rates were often kept secret by the master mechanics," writes labor historian Walter Galenson, "to prevent their employees, who were usually paid by the day, from knowing what their profits were."[6]

Strains appeared in the assumption of common interests between masters/journeymen/apprentices. Master carpenters increasingly defined themselves as employers. In a 1791 account, the masters described their role as "procuring materials, superintending the workmen, and giving directions," as well as "providing tools for the different kinds of work and shops in which it may conveniently be performed."[7] To the degree that the master acted as a contractor, journeymen and apprentices inevitably responded as employees. In 1825, close to six hundred Boston house carpenters went on strike for a ten-hour workday. The masters, not ready to play the part of employer to the hilt, expressed "surprize and regret" at the journeymen's actions. In rejecting shorter hours, they claimed to be baffled by the strikers' demand: "Journeymen of good character and of skill, may expect very soon to become masters, and like us the employers of others; and by the measure which they are now inclined to adopt, they will entail upon themselves the inconvenience to which they seem desirous that we should now be exposed!"[8]

The puzzled master carpenters did not fight the strike single-handedly. Boston's merchants and capitalists hired the masters to build their expanding capital stock. This powerful group of men—about to reveal itself as the pivotal force in the industry—saw the strike in a larger context, as a potentially infectious virus that might inspire other Boston-area workers. These "gentlemen engaged in building" (as they styled themselves) shared none of the masters' confusion. They warned that a successful strike would surely "extend to and embrace all the Working Classes . . . thereby effecting a most injurious change in all the modes of business."[9] At a meeting on the evening of April 21, they agreed to refuse work to any striking journeyman or any master who accepted the strikers' terms.

The spirit of rebellion was in the air. After all, these were the same Boston carpenters who, a generation earlier, were credited with the idea of dumping the tea in Boston Harbor. In 1832, the ship and house carpenters of Boston and Charlestown tried again. On May 23, they drafted a statement outlining their grievances. We work from sunrise to sunset for an average of $1 a day, the carpenters complained, while losing one-third of a year's work due to foul weather. From that day on, they announced they would work only ten-hour days from March until September. This time the merchants and ship owners were even better organized. They declared a lockout of the strikers and raised the staggering sum of $20,000, which they made available to the masters. Accordingly, the master ship carpenters were able to offer $2 a day (*twice* the normal wage) to forty journeymen who would break the strike.[10]

The persistence of Boston's carpenters fueled a general labor uprising in the 1830s. The movement for a ten-hour day swept across New England to New York and Philadelphia. Between 1833 and 1837, building trades workers went on strike thirty-four times, and almost three hundred thousand wage earners of all kinds joined trade unions or mechanics' associations. Seth Luther, Boston house carpenter, won nationwide fame as a labor reformer with the distribution of his *Address to the Working Men of New England*. Luther signed the circular of Boston's carpenters, masons, and stone cutters when the city's tradesmen struck for the ten-hour day once more in 1835. Yet, despite all the labor activity, the results were familiar. After seven long months, the Boston strike collapsed.

The strikes of 1825, 1832, and 1835 failed to shorten the workday, but they did succeed in changing the carpenter's view of his industry. The protests had exposed the shifting relations between working carpenters, masters, and merchant capitalists. The bond between masters and journeymen, once taken for granted, had been frayed by the political alliances formed by masters and merchants in each of the three strikes. In the pre–Civil War United States, apprentices and journeymen could still reasonably expect to eventually become, and could therefore identify with, masters, but the actors in the building industry were leaving the guild traditions behind and entering a new phase. As the 1835 circular put it, "We should not be too severe on our employers, they are slaves to the Capitalists, as we are to them. The power behind their throne is greater than the throne itself."[11]

———

The Panic of 1837 crushed the labor organizations and economic aspirations of Massachusetts carpenters, but the following decade brought another burst of building activity across the state. In Newburyport alone, 832 new homes were built in the late 1840s on top of a previously existing stock of just 600 houses.[12] The boom-and-bust pattern continued for the rest of the century. From 1866 to 1906, the volume of building activity shot up by 250 percent, an upsurge that included wild swings of unbridled expansion and sudden crashes. The pace of new construction was part of a country in change. Railroads connected once remote towns and villages, mechanical inventions wiped out entire handicrafts and extended the factory system to new industries, as the United States traveled from a sea of self-contained communities to a unified nation linked by modern communication and transportation systems.

Carpenters in Massachusetts were respected members of their communities in the mid-1800s. Their work was a "calling," not just a job, and their pay rates reflected this position. Figures from Essex County in 1875 show that though the average carpenter worked only slightly more than seven months a year, his annual income was 41 percent higher than that of factory workers and 49 percent higher than that of common laborers. In fact, among skilled craftsmen, only blacksmiths and machinists received more—and they worked year-round. The wages secured the carpenter's social status in his community. Unlike the dependents of a semiskilled or unskilled workman, writes historian Stephan Thernstrom, the artisan's "wife and children were under much less pressure to enter the labor market themselves to supplement the family income."[13]

Along with the demand for construction skills, the Industrial Revolution brought some troubling problems for the carpenter. By and large, the work had not changed since the 1700s. If anything, the knowledge required had been complicated by the growing number of wealthy homeowners whose tastes expanded from the spartan designs of the colonial era to the elaborate detail work of Victorian buildings. In any case, from stark warehouse to lavish mansion, the carpenter still fabricated doors, windows, stairs, trim, and mantels by hand.

The technological innovations that had transformed the textile and shoe industries were eventually applied to the building trades. Steam power and the circular saw had mechanized wood-cutting to uniform sizes by 1815, but at that time all the dressing and fitting of the lumber was still done on the building site or in the shop. After 1840, however, the floodgates of change opened. Factories with new machines—cut-off saws, mortising and tenoning machines, borers, compound carvers, and power sanders—revolutionized woodworking. For a number of years, hand labor and machine production existed side by side. In 1872, John Richards, chronicler of the industry, reported: "It has been an even race, to say the least."[14] By the end of the century, the race was over, and hand millwork had been left in the dust.

In *Empire in Wood,* an excellent history of the Carpenters Union, Robert Christie compared hand- and machine-produced blinds, doors, ceiling boards, flooring, and stairs during the period from 1848 to 1896. He found that machines reduced production costs by 90 to 97 percent![15] With this incentive, entrepreneurs lost little time investing in planing mills, sawmills, and furniture factories. New York State had fifty-eight planing mills worth $131,000 in 1850; fifty years later it was a $22-million industry.[16]

Along with the menace of the factory, the Industrial Revolution spun off a threatening new family of building materials. Cast-iron replaced wooden beams beginning in 1852, and by the turn of the century, the inventions of structural steel, reinforced concrete, and the elevator laid the groundwork for the modern skyscraper. In 1906, a building in Bridgeport, Connecticut, was finished without a stick of wood in it. As wood-free city skylines became more common, one carpenter lamented, "When one looks at the enormous amount of iron, steel and other hard materials which enter the construction of modern buildings, I am tempted to ask . . . what sort of a future this trade has before it."[17]

The industry was changing in other ways.

The opportunities for profit generated by economic growth attracted a group of investors who knew nothing of the building process. Whether they had access to large capital reserves or simply managed to get a ninety-day builder's loan, they pushed the masters aside and stepped in as the new employer. This new type of contractor, sometimes a former master but more often a ward politician or a speculator, knew little of craft pride and the remaining guild traditions of loyalty between masters, journeymen, and apprentices.

Unable to supervise the work directly, these contractors turned to the system of "lumping." The lumper, as labor journalist John Swinton described it, "takes a whole job at a certain figure; he then sublets it to another, who, in turn, parcels it out to others, who do the work in as rapid . . . a manner as possible—tearing and rushing to get it done. They all have to make a profit, at the expense of the buyer and the laborer."[18] The lumper worked piece-rate and usually focused on one aspect of the trade. Specialization became sufficiently common that, shortly after the Civil War, a carpenter reported to the Massachusetts Bureau of the Statistics of Labor that the all-around carpenter had given way to four distinct trades—stair building, framing, finishing, or sash, blind, and door making.[19]

The new breed of contractor emphasized speed and production. John Bicknell, secretary of the Amalgamated Building Trades Council of Boston, estimated that building tradesmen were expected to double their past output.[20] The rapid changes severed personal connections between builders and employees. At one time, a carpenter might have worked for a single master for twenty or thirty years; by the late nineteenth century, he might have as many as twenty or thirty employers in one year.[21]

At the turn of the century, a Connecticut carpenter named J. W. Brown looked back at the previous five decades of his trade. In the past, he wrote, "when an employer hired a carpenter, he would, as a general rule, send his team after his [tool] chest and have it taken to the shop. And as long as the man worked for him the employer felt himself under a moral obligation to keep him employed steadily, and when a rainy day came he always provided for him in the shop or elsewhere." Today, Brown continued, all the carpenter needs is a "collar box" of tools, which goes to the job not in the employer's team but on the carpenter's back. Arriving on the site, the carpenter does as he is told—putting a floor on some brick block or nailing up factory-fitted trim. He spends a few weeks "picking at such a job," packs up his tools, "and takes up his march again."

Brown recognized the carpenter he described as a new kind of tradesman, one who had lost a sense of certainty that he would become a master and carry on the guild heritage. He also knew that the volcanic Industrial Revolution had left a changed social system in its wake— the era of modern capitalism, an era of sharply separated social classes whose interests were often in conflict. What was happening to the carpenter was just one chapter in the story of the making of the American working class. For Brown, this meant that the carpenter had become "accustomed to look upon himself not only as a wage worker for life, but as an appendage to a monstrous machine for the production and distribution of wealth."[22]

Building framed in wood, the standard structural method in nineteenth-century Massachusetts. SPNEA

***Working Lives/Union Life:
The Early Years***

Newburyport cabinetmaker. SPNEA

(*Below*) Carpenters resting on mortised beams. SPNEA

A barn raising. IHE–UBCJA

Tradesmen pose in front of their handiwork. Fred Ernest

Boston Public Library construction site, c. 1893. BPL

29

Contractors oversee early stages of building library, 1888. BPL

Placing the cornerstone, 1888. BPL

Wooden form shapes masonry arch in library construction. BPL

A workman inspects the main entrance to the library, c. 1893. BPL

THE
RULES OF WORK,

OF THE

CARPENTERS,

IN

THE TOWN OF

BOSTON.

FORMED, and MOST ACCURATELY CORRECTED, by a LARGE
NUMBER of the FIRST WORKMEN
of the TOWN.

Published agreeably to Act of Congress.

PRINTED,

FOR THE PROPRIETORS.

1800.

This drawing of the first Labor Day parade in New York City on September 5, 1882 appeared in *Frank Leslie's Illustrated Newspaper* for September 16, 1882. The idea for this holiday is generally credited to UBCJA founder Peter J. McGuire. UBCJA

(*Opposite page*) In the early 1800s, carpenters' pay rates and working conditions were governed by local "rules of work." UBCJA

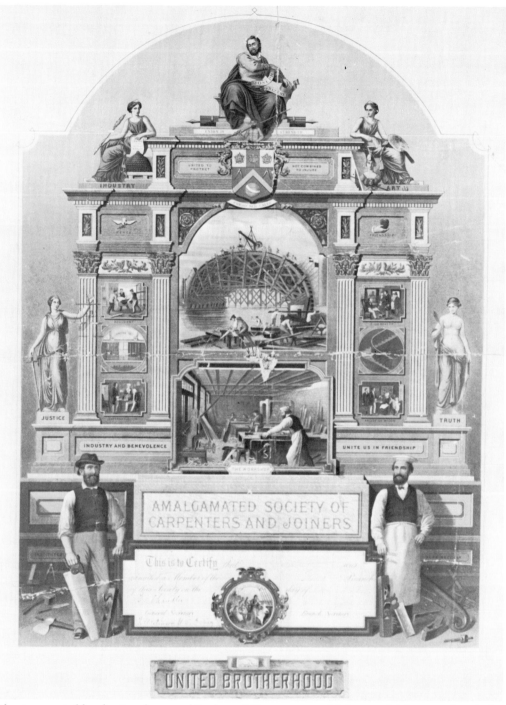

Charter granted by the Amalgamated Society of Carpenters and Joiners, which merged with the United Brotherhood of Carpenters and Joiners of America in 1888, when the words "United Brotherhood" were placed over "Amalgamated Society." UBCJA

34

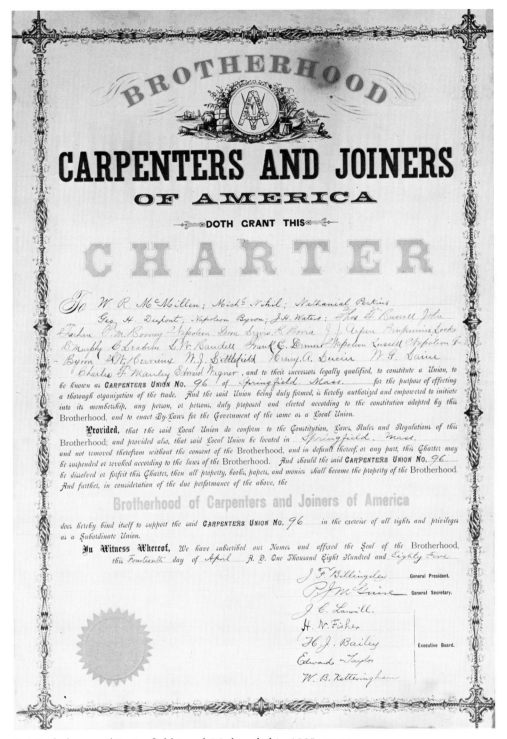

Original charter of Springfield Local 96, founded in 1885. UBCJA

(*Above*) Foreign-language locals, particularly French, played an important role in the early years of the Carpenters Union in Massachusetts. PAC–UBCJA

Local No. 1305 de la Fraternite Unie

DES CHARPENTIERS ET MENUISIERS, D'AMERIQUE.

Cher Monsieur,

Vous êtes respectueusement appelé à assister à l'assemblée mercredi, le 10 Février, pour discuter une affaire importante.

Par ordre de

J. A. COTE, Sec.-Arch.

Notice for a meeting of a French local in Fall River.

Boston Local 33 membership card for 1889–90. SMLA

An Upholder of Cheap Bosses and Low Wages.

HIRAM WORK-AS-LONG-AS-YOU PLEASE.—"These pesky unions tire me.   This is a free country.   A man has a right to work for any price.   If wages are low let us work longer hours to make up."

Cartoon in April 1896 *Carpenter*, UBCJA's monthly journal. UBCJA

Identifying ribbon for convention delegate from New Bedford local. SMLA

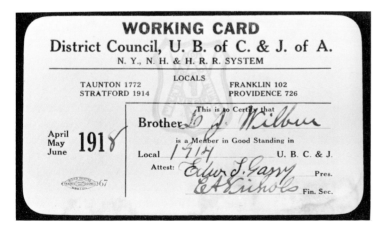

Carpenters carried working cards in their wallets and wore working buttons to indicate that they were union members in good standing. SMLA

4

The Eight-Hour Strikes: 1886 and 1890

There were three hundred thousand J. W. Browns swinging hammers in the United States in 1881. Buffeted by the twin blows of mechanization and a changing industry structure, carpenters were searching for protection in an increasingly harsh and unstable work environment. The terrible depression of 1873–78, the most severe in American history to that time, put an end to remaining illusions of guaranteed security. During those years, contractors imposed whatever wage levels they wanted. Desperate carpenters worked for subsistence pay—if they worked at all. At the height of the depression, a New Yorker asked and answered his own question: "What are our carpenters doing? Nothing! . . . What have they to live on the coming winter? Nothing!"[1]

As the economy recovered at the end of the decade, tradesmen considered the idea of trade unions as shelters against the ravages of the coming industrial order. Carpenters' unions had been formed before, on the local and national levels. They had either failed or turned into unassuming fraternal societies. Peter J. McGuire, a twenty-eight-year-old socialist carpenter from New York, felt the time had arrived for a new initiative. For eight years, he had traveled the country working at his trade in order to support his true

vocation of organizing. In 1881, McGuire was fresh from a successful carpenters' strike in St. Louis. Despite his youth, he had won a national reputation among carpenters. In May, he published the first issue of the *Carpenter,* stating, "In the present age there is no hope for workingmen outside of organization. Without a trades union, the workman meets the employer at a great disadvantage. The capitalist has the advantage of past accumulations; the laborer, unassisted by combination, has not."[2] Three months later, carpenters from eleven cities answered McGuire's call. On the second story of a flaxseed warehouse in Chicago, they deliberated and formed the Brotherhood of Carpenters and Joiners of America (later changed to the United Brotherhood of Carpenters and Joiners of America).

None of the delegates at the Chicago convention came from Massachusetts. Nonetheless, the idea of a union caught on very quickly among the commonwealth's carpenters. The first charter for a local union in Massachusetts was granted to a group of carpenters in Holyoke on March 1, 1882. Holyoke Local 455 was soon followed by Local 33 in Boston. In the next seven years, local unions were founded in Cambridge, East Boston, Fall

40

River, Haverhill, Hingham, Hudson, Lawrence, Lynn, Natick, Newton, Quincy, Roxbury, Springfield, and Worcester.

In some cases, it was a matter of a few interested carpenters applying for a charter. Local 192 in Natick was organized in 1886 with only 14 members. Quincy L. U. 417 started two years later with 26 members. Some locals dissolved after a few years; others grew quickly. Local 417, for example, had a hundred members by 1895, representing about 65 percent of all the carpenters in town. Lynn Local 108, with an initial membership of 116 in 1889, claimed 636 members only four years later, which, according to the local, included "nearly every journeyman carpenter in the city."[3] Boston was the hub of union activity. Local 33 had 404 members in 1883 and doubled in size in the following ten years. Boston carpenters had the largest locals, the highest wages, and usually set the tone for union campaigns.

The shorter workday was the most pressing issue for Massachusetts' pioneering union carpenters. By 1880, the ten-hour day was the industry standard. Tradesmen now sought to restrict the working day to eight hours to allow them more time for their families and the chance for "self-improvement." This was a powerful attraction for nineteenth-century workers who had limited access to schools, libraries, and museums. The shorter day represented a form of liberation, an opportunity to enjoy and make something of the precious leisure time of life. Eight-hour-day advocates freely borrowed from the language of the Revolution, equating their campaign with a fight for freedom from tyranny. They drew a sharp distinction between "their time" (the eight hours belonging to the employer) and "our time." The eight-hour movement had its own cultural trappings and symbols; people smoked "Eight-Hour Tobacco," wore "Eight-Hour Shoes," and sang the "Eight-Hour Song"— "Eight hours for work, eight hours for rest, and eight hours for what we will."

The movement for shorter hours peaked in 1886, a year often referred to as "the great uprising of labor." Never before had so many American workers acted in unison for a common goal. There were more than seven hundred strikes in Massachusetts in 1886—more than *seven* times the number of strikes during the previous two years. Across the country, 340,000 workers demonstrated for the eight-hour day. Scanning the national scene, the Wisconsin commissioner of labor wrote: "The agitation permeated our entire social atmosphere . . . it was *the* topic of conversation in the shop, on the street, at the family table, at the bar, in the counting rooms, and the subject of numerous able sermons from the pulpit."[4]

The Brotherhood was in the forefront of the campaign. During the late winter and early spring, Boston's Local 33 sponsored rallies, building support and enthusiasm for the eight-hour cause. In mid-April, the city's carpenters, along with other tradesmen, formally presented their demands for the shorter workday and a new wage scale to the recently organized Master Builders Association (MBA). When the Association refused to negotiate, the carpenters declared that "the workingman's hour" had arrived and announced plans for a strike.[5]

On May 1, seven thousand carpenters, painters, and plumbers shut down Boston's construction sites. Their tactics were simple. As long as craftsmen from neighboring towns and states could be kept out of Boston, the employers would be cut off from an adequate supply of skilled labor. Local 33 immediately sent notices to Maine and the Canadian Maritime Provinces urging carpenters to stay away. Local unions in Somerville and Chelsea guaranteed that their members would not scab; Cambridge carpenters went one step further, pledging one-third of their pay if needed. Carpenters working for the McNeil Brothers struck their jobs in Lenox and Newport, Rhode Island, to prevent the firm from transferring them to its Boston sites.[6]

Other workers—clothing workers, tailors, and cigar makers—joined the strike, virtually paralyzing the city. Nonstriking workers provided additional support. Railway employees ripped down contractors' advertisements for strikebreakers and warned the striking carpenters whenever a trainload of scabs pulled in. But it was the carpenters themselves—often picketing job sites up to sixteen hours a day—that made the strike. They watched the night trains and the boats from Nova Scotia. On a typical day during the walkout, one group of strikers stopped six strikebreakers on the St. John's steamer while another sneaked on board the *Pavonia* and "with an instinct of which only carpenters are capable, spied one solitary Swedish carpenter."[7] The unions never stopped organizing. At a mass meeting on May 7, two hundred nonunion men were sworn in. Even the strikebreakers were treated generously, usually offered the choice of union membership or train fare back home.[8]

The masters were not idle. Lacking the unionists' public support, they consolidated their forces. While the carpenters of Local 33 and their allies in the Amalgamated Society of Carpenters and Joiners met in crowded workingmen's halls around the city, the MBA assembled at the plush Parker House. Their position was uncompromising. MBA Secretary William Sayward announced that the employers were ready to put up with the strike for the entire summer before they would permit "conferences or agreements with societies that thrust themselves in between the workmen and the employer." Benjamin Whitcomb, president of the MBA, was blunter: "We can stand it better than they can."[9]

The carpenters' vigilance extended into the second week of the strike. They picketed by day and plotted strategy at nightly meetings. On May 10, 225 picketers persuaded thirty men to leave a job while another group worked the docks, turning back strikebreakers from Yarmouth, Nova Scotia, and Savannah, Georgia.[10] Weekly mass meetings at the Columbia rink drew up to three thousand strikers, but eventually the MBA's stubborn stance took its toll on the workers' morale.

Though a few of the city's small contractors

granted the eight-hour day, the unions found it difficult to crack the larger builders' unity. When Norcross Brothers agreed to work nine hours, the MBA quickly forced the straying firm to retract its offer. On May 17, the unions lowered their demand from an eight- to a nine-hour day. The MBA refused to consider the union's new position. Sensing victory, MBA spokesmen announced that building employers would sooner go out of business than sign any agreement.[11]

Over the next few days, union leaders abandoned the battle. Concluding that the membership was not prepared for the lengthy strike it would take to overcome the MBA, they admitted defeat at a May 20 mass meeting. The decision shocked the crowd of two thousand. Cries of "no, no," "we won't do it," and "don't give in" echoed around the rink. As speaker after speaker recommended a return to work, the strikers reluctantly accepted their verdict.[12]

Many of the speeches that night were bitter. George Rothwell, president of the carpenters' executive committee, attacked the arrogance of the contractors. "You have lost the confidence of the men who worked for you," he charged. But others, like A. A. Carlton of the Knights of Labor and venerable labor leader George McNeill, argued that the strikers' show of strength had brought a moral victory. "This was not a strike but only a notice of what is to come," argued McNeill. "We go back to take in reinforcements and supplies."[13]

The final statement by the strikers, adopted overwhelmingly at the meeting, reflected both anger and hope for the future. Accusing the contractors of "galling abuses," the resolution declared that the very term "master" in the title of master builder was "foreign and offensive to our sense of citizenship." The carpenters also had some harsh words for the steady stream of strikebreakers, whom they labeled "a curse to humanity and a blight to their fellow craftsmen."[14]

Finally, the strikers called for improved organization to stop, once and for all, the MBA from "stamp[ing] their heels on our heads and eat[ing] the whole oyster of our labor while they throw to us the shells."[15] The seventeen-day strike had brought hardship and no victory. It was small consolation—but the only one they had—for the strikers to know they had fought against long odds.

━━━━━━━━━

In 1886 they met a rabble. In 1890 they met a band of drilled and tried trade unionists.
— Harry Lloyd, vice-president Local 33[16]

The commitment and sacrifices of workers throughout the nation in 1886 stunned the business world and surprised cautious union officials. Thousands of rallies, walkouts, and strikes—many of them spur-of-the-moment actions—demonstrated the powerful appeal of the eight-hour day. Surging rank-and-file sentiment persuaded labor leaders to prepare for another bout with employers. Planning was the key. The spontaneous nature of many of the protests had shaken up whole cities, but it had not necessarily brought victory.

Delegates to the 1888 American Federation of Labor (AFL) convention voted in favor of a coordinated series of actions for May 1, 1890, and instructed the executive board to devise the most effective strategy. The board elected to commit the Federation's resources to the single strongest union instead of sponsoring a number of scattershot demonstrations. The obvious candidate was the United Brotherhood of Carpenters and Joiners of America (UBCJA), an organization described by AFL head Samuel Gompers as the "best disciplined, prepared, and determined" force in the labor movement.[17]

Gompers's judgment was accurate. The UBCJA had thirty-two thousand members in 1888 and an annual budget of $67,000. The union had 464 locals representing every state of the union (including 40 in Massachusetts).[18] Presented with the responsibility of leading the fight for the entire labor movement, General Secretary McGuire and the other national officers granted locals in a dozen cities use of the Brotherhood's sizable strike fund. Though carpenters ultimately struck in 141 cities, it was these union strongholds that carried the brunt of the battle.

Tension in the Boston industry had not eased since the 1886 walkout. The MBA had instituted a nine-hour day the following year "in order to rationally test the question of whether ten hours per day is too long for men to labor." But this carrot had been accompanied by a stick. The builders warned that if workers chose to strike, "then we shall feel at liberty to return at once to the old standard of ten hours." The MBA also drew up a Code of Working Principles that insisted on complete employer "control and authority." Under the code, contractors promised to fire tradesmen who refused to work with a nonunion employee or install materials without a union label. They rejected on-the-job organizing pressures and emphatically declared that "no person outside the employment of said contractor shall be allowed to represent [the employees]."[19]

The MBA had not wavered in its hostility to unionism. By 1890, carpenters were eager to rearm for combat. Boston's carpenters had chartered additional locals in Roxbury, Dorchester, Charlestown, East Boston, and Brighton. Once again, they set the wheels of protest in motion. This time, however, unions in the rest of the state were not content to stand back and provide support for their big-city brethren.

Carpenters in Lynn fired the first shot, refusing to work past 5 P.M. after April 1. Almost immediately, the contractors agreed, granting the nine-hour day plus a 25- to 50-cent-a-day wage increase. Eight-hour meetings were held in Fitchburg, Lawrence, Milford, Salem, and Taunton. Locals in Gloucester, Haverhill, Malden, Leominster, Lowell, and Waltham all won a nine-hour day after briefly going out or threatening to strike. In Springfield, eight hundred carpenters marched on April 17 and heard George McNeill predict that when the carpenters won, the other trades would fall in line "one by one," bringing about a universal eight-hour day. In Worcester, twenty-five hundred tradesmen rallied at Mechanic's Hall on April 18, preparing to join the Boston carpenters for a May 1 strike.[20]

In Boston, the unions spent April readying the membership and garnering support from outside sources. Every church in the city received a letter requesting at least one pro-eight-hour sermon from the pulpit before the end of the month. Union meetings reached new attendance heights. Overflowing into stairwells and waiting rooms, meetings lasted far into the night as dozens of new members were initiated and old-timers debated strike plans. As organizing reached a fever pitch, the *Boston Globe* editorialized that the city could expect "the greatest labor demonstration ever seen."[21]

The employers had made their own preparations. Since 1886, the Master Builders Association had spun off organizations for each branch of the industry, ready to take on their own group of tradesmen. The Carpenters Builders Association (CBA) informed its employees that it had no intention of discussing demands for a shorter day. All of the unions' carefully arranged plans were suddenly jeopardized when Norcross Brothers locked out their freestone cutters in mid-April. In the face of mounting labor militance, the MBA had decided to, as one contractor said, "take the bull by the horns" and raise the specter of an industrywide lockout. Norcross had adopted a tough antilabor policy since their moment of weakness in 1886. Their action against the freestone cutters, a small and highly skilled craft that had won the eight-hour day, was a challenge to the other trades' resolve. Orlando Norcross stated that he would rather fight the trade un-

ion movement "forever" than make any concessions.[22]

The lockout had its intended effect, throwing the building trades into an uproar. Carpenters walked off Norcross's Ames Building job in sympathy, only to return at the direction of the union officers. Masons refused to follow their leaders' orders for a similar action at the Norcross State Street Stock Exchange job because the carpenters continued to work. By the time the Amalgamated Building Trades Council convened on April 13, the fissures in trade harmony were opening. Meeting in crowded Pythias Hall, the delegates ordered each union's walking delegate or executive officer to call his tradesmen off every Norcross site in New England. "This is our answer," said one unionist, "to the declaration of war on the part of the employers." Those who spoke in favor urged the combined trades to drop the May 1 deadline and join the carpenters in an immediate industrywide strike for the eight-hour day.[23]

Though the carpenter delegates voted for the Pythias Hall strike call, Local 33 officials reconsidered in the next few days. Fearing that a premature walkout would disrupt the original eight-hour strategy and jeopardize national union support, Walking Delegate Joseph Clinkard refused to pull carpenters off Norcross jobs. The bricklayers were also unwilling to participate. Without the support of the two largest crafts, the Building Trades Council's war plans fizzled. The Brotherhood's locals returned to their original plans, finalizing strike arrangements during the last two weeks of April. But with the Norcross lockout, the MBA had exhibited potent weapons in its arsenal.[24]

As the sun rose on the morning of May 1, one hundred small building contractors put two thousand carpenters to work on an eight-hour basis. The larger member firms of the Carpenters Builders Association refused to negotiate, but it seemed only a matter of time before their eighteen hundred employees would have the shorter day as well. The strikers were well organized and had the financial backing of the national union, other AFL unions, and an assessment on the wages of those union carpenters working for non-CBA employers.

Bulletin boards at strike headquarters at the Tremont Temple directed members to their picket duty. Strikers who answered morning roll call and picketed daily were entitled to strike pay—$5 a week for married men; $4 a week for unmarried men. Fifteen hundred turned up at Tremont Temple the first day.[25] Spirits were high at a mass meeting that night in a hall "packed to suffocation." Picketers reported on the day's successes to cheers of approval. A week later, the mood was unchanged. John White, president of Local 33, boasted: "It will only be a short time when we shall have eight hours and we can say, 'we are the people.'"[26]

As in 1886, stopping the strikebreakers was the key to victory. The Brotherhood had expanded across the border and their new Canadian locals acted to cut down the influx of carpenters from Quebec and the Maritimes. Knowing that, the contractors had redoubled their recruiting efforts, focusing on rural areas that were freer of union influence. Every day union scouts patrolled the docks and depots; every day pickets checked the job sites.

May 2. Chief of pickets reports that picketers convinced forty-eight more men to walk off their jobs. Contractors forced to offer extra pay to those who remained.

May 4. Twenty-five hundred carpenters out of a total of thirty-eight hundred now are working an eight-hour day.

May 5. Strikers notice carpenters' tool chests on board a Nova Scotia steamer. Committee of strikers dispatched to Commercial Wharf. Soon after, they return to union hall having signed up eight strikebreakers "amid tremendous cheers from the large gathering in the hall." Other nonunion carpenters, re-

cruited by CBA employers, join the union when informed of the strike situation. Some of the smaller CBA contractors begin to break away. Three hundred more granted eight hours.

May 6. Clinkard claims the availability of strikebreakers is so low that a contractor who had advertised for scabs was forced to look for workers at the strikers' headquarters. A newspaper reports that another firm was so desperate for help that it bailed a man out of the Deer Island prison. Despite the strike's effectiveness, the CBA refuses to negotiate.

May 8. Daily picket report claims sixty more men pulled off jobs.

May 9. Local 33 leader William J. Shields returns from the national union office in Philadelphia with strike funds. He describes huge membership gains across the country and reports the "money is rolling in." The strike is now a "question of endurance," he claims.

May 12. Fifty-five more nonunion carpenters walk off a job, and twenty new members are initiated into Local 33. Another member of the CBA breaks ranks, declaring he could no longer stand "being dictated to by men who had accumulated fortunes out of their carpenters and who were too stubborn to grant a fair request." One of his single employees, now able to return to work, offers his job to a married union brother who needs it more.

May 13. Reports of dissension continue to filter out of the CBA. "Some of the smaller builders are beginning to grumble . . . and say that the big fellows are eating them up."

May 15. Seven more strikebreakers are confronted at the Nova Scotia boat and are persuaded to join the union.

May 18. Non-CBA contractors request sixty-five strikers to work at eight hours, but many refused the work "so desirous were they to remain in the fight."[27]

Throughout the month, the leaders of the CBA ignored the unions' demands and tried, as best they could, to fill their jobs. Despite their exhausting efforts, the unions simply did not have the resources to stem permanently the steady tide of strikebreakers. By mid-June, the union was forced to admit that most job sites were staffed.[28] Even with the added pressure of a walkout of bricklayers and building laborers, the CBA maintained its unyielding posture. Responding to an offer of mediation from the State Board of Arbitration, E. Noyes Whitcomb, president of the CBA, wrote a curt three-line note on June 23 dismissing the state's interference. In its subsequent report, the Board noted ruefully that its efforts had "no perceptible influence" on the conflict. The union had welcomed the Board's attempts, and its failure deflated the carpenters' hopes of victory. In fact, soon after Whitcomb's rejection the strike "began to show signs of dissolution."[29]

On July 12, the strike committee advertised the availability of carpenters for hire. The $5 weekly strike benefit was no longer enough; the men were feeling the pressure of two and a half months without a pay envelope. The financial strain of strike support was also taking a toll on the national union. When the Boston District Council complained of inadequate help, the General Executive Board in Philadelphia tersely pointed out they had already sent $9,600. The Board went on to ask, sarcastically, who had "ordered them out on strike."[30] The strikers knew time was running out. Carpenters in thirty-six other cities had won the eight-hour day, but Boston's large employers showed no signs of bending.

Breaking with the open character of the strike, all the Boston locals met secretly on July 23. At a mass rally the following night, union officials confessed their inability to continue support to the struggling strikers. The strike was not called off, but members were free to go back to work. The union tried to make the best of a deteriorating situation. Spokesmen transformed the mass walkout into a "harassing plan of guerrilla warfare" in which "no contractor will know when his job will be struck

and all his operations impeded." The plan proved ineffective. A carpenter's banner in Boston's massive 1890 Labor Day parade reflected the unchanged hours of work, declaring eight hours was still "our future time."[31]

Worcester's carpenters had no more luck than their Boston brothers. They had decided to delay a May 1 strike hoping their contractors would voluntarily grant nine hours. When the employers refused, one thousand union and two hundred nonunion carpenters laid down their tools on June 23.[32] The strike paralleled the course of the Boston walkout. Several small contractors accepted the new schedule, but the big builders, again led by Norcross, stood firm. "I shall sign no union agreement," proclaimed James McDermott, "for I have nothing to do with any union as I run my business myself." J. G. Vaudreuil was no more conciliatory. "The only way to settle the strike is for the strikers to go to work."[33]

Hundreds of picketers patrolled construction sites and guarded the Union Railroad Station. Dozens of Canadian and New Hampshire strikebreakers were turned away, but contractors managed to slip others by, particularly "P.E.I. [Prince Edward Island] and Nova Scotia men," according to one striker. For a few weeks, the picket lines held firm and the union gained considerable public support. When the contractors asked for legal restraints on the picketers, the Worcester police refused to make arrests.[34]

Boston's protest had been consistently peaceful. Worcester's carpenters showed less reluctance to take matters into their own hands. Picketers attacked strikebreakers and scattered tools on job sites. Local 93's leader C. D. Macomber spent much of his time bailing men out of jail and paying fines on trespassing and other charges.[35]

A shortage of money proved fatal. The national union made a special $1,000 contribution to Worcester, but the rest of the Brotherhood's relief funds had been earmarked for other cities. In fact, McGuire was outraged when the Worcester local sent out a nationwide appeal for aid implying authorization by the General Executive Board. He admired their firmness, he wrote in the *Carpenter,* and he would not discourage any financial support. But he took special pains to point out that help for Worcester was "entirely optional."[36] After eight weeks, the strikers returned to work on a ten-hour basis.

Both the Boston and Worcester strikes had been impressive demonstrations of the rank-and-file's staying power and an affirmation of principle in the workers' willingness to sacrifice an hour's pay for the sake of an idea. George McNeill praised the Boston strikers for their unprecedented action, "the first time in history . . . that workers voluntarily asked to reduce the hours of labor at their own expense." That the city's carpenters fell short of their goal was testimony not to their failings but to the intransigence of the city's building employers. In 1886, MBA chief William Sayward (who, by 1890, was head of the National Association of Builders as well) stated that the conflict involved far more than "the superficial question of eight hours." The union agreed, noting that the work stoppage was "not so much a matter of hours and wages as a matter of recognition." In other cities contractors had granted shorter hours with little fuss. In Massachusetts, Sayward and his colleagues rejected the union's right to speak for their employees on *any* matter concerning working conditions. They accepted a summer of economic chaos rather than lose a battle of principle.[37]

5

Union Building

The crowd quieted as Local 33's Harry Lloyd stepped to the podium of Faneuil Hall. As chairman of the giant March 20, 1894, labor rally, Lloyd opened the meeting. He introduced the speakers, including featured guest Samuel Gompers, and set the tone for the evening. "During the last winter," Lloyd began, "we have had a depression such as our country has never seen . . . our members were on the verge of starvation."[1]

The panic of 1893 had, as Lloyd observed, closed the doors of hundreds of banks and thousands of business firms. At the peak of the crisis, an estimated fifty thousand in Boston and two thousand in Springfield were out of work. In smaller towns, unemployment rates were irrelevant. There was simply, as the *Boston Globe* put it, "no work." The panic extended the anxiety of joblessness to a previously untouched spectrum of the population. Itinerant workers had become a permanent feature of the American economic landscape but the winter of 1893 witnessed a new and "better class of tramps." A study of homeless workers in fourteen cities revealed that over half had "trades, employments or professions requiring more or less skill." Men were tramping who had never tramped before.[2]

Working-class hardships intensified class tensions. The usually cautious Lloyd lashed out at the "kid glove aristocracy" of the city of Boston where the gulf between the rich and poor was particularly visible and irritating. In his Faneuil Hall speech, Lloyd contrasted the "miserable pittance" of charitable donations for the unemployed with the $400,000 worth of contributions to the new music hall. The wealthy "listened to the singers while the poor starved," he commented bitterly.

Editorializing in Boston's *Labor Leader*, Frank Foster angrily charged "that those who built the palaces of the Back Bay should not be compelled to live in the tenement houses of the slums." Unfortunately, the palace builders were stuck in their tenements with little chance of a regular job. The Massachusetts Bureau of Labor Statistics reported that 10,747 of the state's 22,781 carpenters were out of work that year.[3]

The depression set back the development of the Carpenters Union. Membership dropped as carpenters strained to pay union dues, let alone food and shelter bills. Two-thirds of the national union's members were unemployed and over a third of the locals suffered wage reductions. Boston's carpenters managed to hold the line on wages. Local 56 turned aside a contractor offer of steady employment in exchange for lighter pay envelopes. As carpenters lost jobs and bargaining leverage, the number of strikes declined. Throughout the crisis, P. J. McGuire preached against hasty strikes. He argued that any ill-advised job actions, however justified, were likely to be defeated and would add to demoralization in the ranks. The 1894 national convention acted to ensure discretion, limiting the number of officially sanctioned strikes and outlawing winter work stoppages altogether.[4]

Hard times increased the appeal of a shorter day. With wage increases out of the question, carpenters once again focused on reducing the hours of work. By 1892, carpenters in forty-six cities worked for eight hours, but none of those cities was in Massachusetts. Veterans of the 1886 and 1890 Boston strikes publicly criticized those who had stood on the sidelines. They argued that a victory then might have helped cushion the misery of widespread unemployment in 1893. The unemployed were clutching at straws; they welcomed any proposal to spread the available work around. The Master Carpenters Association of Boston agreed to begin an eight-hour schedule as an emergency measure by November 1893, only to renege on their promise in October.

Working-class fortunes picked up in the second half of the 1890s. Economic recovery settled the panicked employment environment and encouraged carpenters to raise their banners of 1886 and 1890 again. In June of 1894, a New England convention of union carpenters discussed a coordinated eight-hour-day campaign, but it was left to the tightly organized Lynn Local 108 to crack the eight-hour barrier in November of 1894. The Lynn local stands out as a particularly militant and effective union in the 1890s, in large part due to the city's long labor tradition and the prominent support of the well-organized shoe workers' unions. Through the rest of the nineties, other locals followed Lynn's example, exchanging wage freezes or even reductions when necessary. By 1902, employers in virtually every town in the state accepted the shorter day as the industry standard.

The national union more than made up its depression losses, counting 68,463 members in 1900. Four years later, the state of Massachusetts alone had 12,000 union carpenters. From the end of the depression until World War I, carpenters in Massachusetts assumed more and more control over their working lives. Union membership grew steadily, and carefully planned strikes brought on the closed shop and an elaborately constructed system of work rules designed to protect and advance the interests of the working carpenter on the job site. In many ways, this was the most significant union-building era in the history of the Brotherhood, as it consolidated its position in every city and town in the commonwealth.

Union members shed their depression-era mentality and resumed their posture of militancy. Demanding better wages, shorter hours, and the enforcement of union work rules, Massachusetts carpenters laid down their tools with increasing frequency. From January to December of 1900, they called twenty-one strikes across the state. Hard times had taught carpenters to choose their battles carefully. Fifteen of the twenty-one strikes succeeded either partially or completely. The results of four were unclear and only two failed outright.[5]

The high success rate prompted many employers to avoid conflict. Typically, a representative of the union approached local contractors with a request for a wage increase and an eight-hour day. After a few quick conferences, a settlement would be reached based on the original union demand or a slightly modified version. In those cases when employers were reluctant to comply, a short walkout was often enough to change their minds. In Springfield, for example, carpenters asked for the eight-hour day with no loss in pay ($2.75 a day) in 1901. The masters met, discussed, and rejected the union proposal. But further negotiations split the masters. Some granted the shorter day; others capitulated after several days without employees. As a contemporary account reports, "The victory was so substantial that a closed shop agreement was entered into the same year."[6]

The issue of the closed shop, that is, the exclusive hiring of union members, ranked as high as wages and hours as a priority for union carpenters. The presence of nonunion workers on a job was a sore spot, a reminder of incomplete union authority. A thoroughly organized workforce was a precondition to job-site control. Even a handful of men outside the union community represented a threat, encouraging employer disdain for union work rules. The Lynn local had won preferential hiring for union members along with the eight-hour day in 1894. Locals without such agreements were forced to monitor and challenge employer hiring practices constantly. In 1898, for example, Springfield carpenters refused to stay on a L. F. Carr building project. The carpenter foreman had declared an open shop policy, remarking that union membership was not "any of my business any more than what church they belong to." The strikers accused him of abusing the workmen and firing without cause. But the real problem, they acknowledged, was the nonunion foreman's stubborn refusal to operate a closed shop. The union carpenters had taken up the grievance "to test their strength."[7]

For some carpenters, mere possession of a union card was not enough. A card-carrying member who violated cardinal principles of unionism was still outside the pale. In May of 1900, seven carpenters employed by Doane & Williams of Holyoke walked off their jobs after the contractor hired a union painter who had worked during a previous building trades lockout. After two days, the Building Trades Council voted that the men could return to work, but only after the Council received a $10 fine from the guilty painter. Union discipline was enforced through strict sets of rules. Brockton Local 624 issued $25 fines to members who worked below the standard rate. The Fall River Carpenters District Council by-laws and the Holyoke Building Trades Council constitution prohibited members from working with nonunion men. These rules were obeyed. A Holyoke contractor named Dibble had hired sixty-five union carpenters and three union painters on a project for the local Street Railway Company. When the railway company decided to bring three of its own nonunion painters on the job, the carpenters' business agent objected. The company brushed off the protest, and all the union painters and carpenters immediately walked off the job. Later that day, a frantic railway representative informed the agent that the painters had been removed and pleaded for the return of the union tradesmen.[8]

During 1901, carpenters staged dozens of successful one-day walkouts against the presence of nonunion men. Carpenters in Fitchburg, Gardner, Milford, Pittsfield, and Westfield prevailed in strikes over wages, hours, and union rules. New locals were formed in Beverly, Newburyport, Northfield, Rockland, Waltham, Wareham, and Whitman. Whether formally acknowledged or not, *de facto* closed shop conditions ruled these carpenters' jobs. There were failed attempts to keep nonunion men off jobs, such as a June 1901 strike by nineteen Cambridge carpenters. But more often than not, the offending parties either joined the union or were dismissed under the gun of a threatened work stoppage. The ultimate sign of the closed shop, of course, was that such actions were usually unnecessary.

Most contractors in Massachusetts hired only union carpenters.

A 1902 agreement in Boston represented an explicit written recognition of the unions' rise to legitimacy. From the early 1880s the fiercely antiunion Master Builders Association had set the tone for the industrial relations in the city. As the unions' permanence became assured with their control over the area's labor market, a growing number of contractors appealed to Association leaders to drop their implacable opposition. Reluctantly, the MBA stepped back from direct involvement in labor relations, leaving those matters to employer associations in each trade. In 1893, the MBA established a Joint Committee for Arbitration that effectively outlawed sympathetic strikes, but administration was left to the separate trades. The Joint Committee broke down ten years later, and the Masters retreated to their downtown offices, petulantly recalling the happier days of 1886 and 1890 when it had not "become the vogue to 'patch up' contentions . . . and to 'recognize' aggregations which had not put themselves into a fit condition for recognition."[9]

Individual craft unions welcomed the diminished role of the MBA. Certainly, the combined power of one set of contractors could not compare with the collective strength of all the city's building employers. Unionized carpenters, in particular, found the Master Carpenters Association (MCA) much more flexible. In March 1902, Boston's five thousand organized carpenters threatened to strike for a wage increase. Two months later, the Master Carpenters granted an eight-hour day at 35 cents an hour. On October 22, the MCA signed a formal contract with the United Carpenters Council, the umbrella organization of the twenty-seven Boston locals. The settlement continued the May agreement on wages and hours, added a 2½-cent raise for 1903, and specified time-and-a-half pay for overtime and double time for work on Sundays and holidays. In addition, the MCA promised to allow union business agents unrestricted access to job stewards during working hours and both sides agreed to submit all disputes to a new joint arbitration committee.[10]

The contract was extended to the eighty-two-member Boston Builders Exchange in April 1903 and governed relations in the industry for several more years. The Masters could never bring themselves to accept a closed shop formally, but in fact, only union carpenters worked their jobs. In turn, union members promised to work exclusively for "recognized builders," that is, members of the MCA or the Exchange. The carpenters' continued demand for preferential hiring for union members was the only unresolved conflict during the years the pact held sway. The original document guaranteed "the principle that absolute personal independence to employ or not to employ" was unquestioned. The wording remained intact. In an arbitration ruling by Judge George Wentworth in 1905, the Council won all its requests (the forty-four hour week during summer months, 41 cents an hour, double-time pay for all overtime, and an extra holiday) *except* the closed shop.[11]

Bargaining successes prompted unions to incorporate further measures aimed at controlling the labor supply. For example, the 1903 Boston agreement established a joint union–management committee for apprenticeship. The unions hoped to resurrect the organized training procedures that had, by and large, suffered in the late 1800s. Indentures continued well into the 1800s, but training was erratically carried out and enforced. Reginald Grover's father moved into a Brockton contractor's home in 1895. Grover remembers hearing that his father's duties included "taking care of the horse and babysitting for the kids."

More often than not, apprenticeship was a low priority for local contractors who were unwilling to make a commitment to an untrained worker in an industry based on short-term employment arrangements. Writing in 1912, James Motley claimed:

The apprentice is simply not wanted: for no large contractor desires boy labor merely because it can be secured at a low wage. The time consumed in teaching the trade to beginners would cause a delay in work, more expensive than the sum saved on wages. Indeed the apprentice is physically in the way, on a large building where all work must be done in order and on time.[12]

Employer indifference proved to be more potent than union interest in the 1903 instance. Five years later, the joint committee collapsed from lack of employer support, an experience that was repeated in cities across the country.

Such defeats, however, were few, and victories were common. The relative strength of union carpenters was by no means matched by other American workers at the turn of the century. On the contrary, the ability of building trades workers to enforce standard wages and hours as well as working conditions made them the exception rather than the rule. The vast majority of the workforce labored outside the trade union movement, and many of those workers who were organized were suffering a state of decline. The number of unionized workers in the expanding manufacturing sector was very small—no more than 6 percent in 1900—and they consisted almost entirely of those highly skilled operatives whose craft knowledge had not yet been undermined by the factory system.

Turn-of-the-century union carpenters were growing in influence and power; factory workers of the same era were struggling to retain what little strength they had. Steelworkers, for example, practically lived in the steel mills, working ten- or twelve-hour days, six and sometimes seven days a week. Most earned between 14 and 25 cents an hour. A survey of twenty-eight steel plants in 1900 showed that only one of every nine steelworkers was paid as much as a carpenter in Springfield. The position of the unions was equally precarious. The once substantial Amalgamated Association of Iron, Steel, and Tin Workers, representing the top strata of the workforce, had fallen victim to a concerted steelmaster antiunion campaign. By 1900, not a single mill of consequence in the steel centers of western Pennsylvania recognized the Association. Regarding workers' ability to press for improved working conditions, F. N. Hoffstot, president of Pressed Steel Car Company, summarized their options neatly: "If a man is dissatisfied, it is his privilege to quit."[13]

How had the United Brotherhood of Carpenters and Joiners of America emerged as the country's largest trade organization when unions like the Amalgamated Association were crushed? With the exception of periods of general economic downturn, how had carpenters been able to improve slowly but steadily their hourly pay and working conditions? And finally, in an era when a U.S. Steel executive could boast: "I have always had one rule—if a workman sticks up his head, hit it," what possessed a man like Otto Eidlitz, one of the nation's most powerful builders, to proclaim: "It is without question, not only the right but the duty of labor to thoroughly organize itself, and it . . . is a power for good in the trade"?[14]

Building in the second half of the nineteenth century was not, as a contractor wrote in 1921, "a gentlemen's vocation."[15] Despite the addition of a number of sophisticated large-scale employers in the 1880s and 1890s, such as Eidlitz in the East, Norcross in New England, Griffiths in Chicago, and Wills & Downey in New York, the industry was still checkered with the local speculator–contractors who had entered the field after the Civil War. The advent of the skyscraper brought forth the professional builder—the executive who scheduled, coordinated, and managed promotion, financing, engineering, materials purchase, and labor relations. But for every such Norcross, Eidlitz, or George Fuller, there were dozens of men who operated by the seats of their pants and carried their offices in their heads.

Low capital requirements encouraged small employers. A minimal supply of tools and equipment combined with a general familiarity with building methods was enough to send many an aspiring contractor off to the construction wars. The irregularity of demand, product, job size, duration, and location offered little incentive for massive investment and the emergence of highly capitalized employers. Furthermore, contractors were only one of the actors in the erection of a building—and not necessarily the major one at that. The owner provided the demand, the banks the credit, the architect the design, the suppliers the materials, and the actual construction was performed by employees of not one but many contractors and subcontractors.

The various parties in construction had differing stakes in the industry's labor relations. Building owners wanted jobs finished and tenants moved in. Delay induced panic and steep financial costs. Owners occasionally threatened to pull contracts if builders ran into on-site labor troubles. Since bank penalties for construction halts exceeded the price of minor advances in worker paychecks, many owners had little patience with virulently antiunion contractors. The contractors, in turn, were willing to forgo endless battles as long as increased labor costs could be passed on easily.

According to builder L. J. Horowitz, the owners' interests tied contractors' hands in relation to the unions. Trade unionism in the building industry was "unavoidable," he argued, "because the average owner is unwilling to have his building operation made the battleground for deciding the issue between a union and

non-union operation." Horowitz contrasted the short-term orientation of construction actors with the well-funded stability of factory owners. An antiunion manufacturer "might justify a battle for two or three years to gain the open shop." That luxury of time was not available to builders. For them, he concluded, "the intelligent course to pursue would be to correct what was wrong in unionism rather than to exterminate it . . . this was clearly a case for a surgeon and not an undertaker."[16]

Even bitter foes of the labor movement like William Sayward reluctantly came to terms with building trades unions. Sayward, the one-time scourge of Boston's carpenters, delivered a speech in 1894 as secretary of the National Association of Builders, admitting that "not only are organizations of workmen right and proper, but they have the elements, if wisely administered, of positive advantage and benefit to the employer."[17] The advantages and benefits referred to by Sayward were the union's ability to solve the industry's chronic problem—a stable source of labor. An employer's need for construction workers ebbed and flowed with the season, total volume of work, and life cycle of a single project. Efficiency and profitability often hinged on absolute flexibility in hiring and firing. In order to compete successfully, contractors wanted the minimum number of workers at any given time to carry out building tasks. As a result, most builders retained a small core of regular employees and recruited the rest on a job-by-job basis.

Since individual contractors could not predict labor requirements with any real degree of accuracy, building trades unions understood the potential power of a union-based hiring hall. A local that contained the majority of a community's construction labor pool became an invaluable asset. If a union could supply half a dozen skilled craftsmen on a day's notice, the employer was relieved of a lengthy, expensive, and unreliable recruitment process. This solution to the labor supply problem far outweighed the added costs to builders of union recognition and above-average construction wages.

Building employers understood the pitfalls of this bargain. Ceding control over labor supply patterns represented a significant loss of managerial authority. They periodically attempted to correct their vulnerability. In 1888, the National Association of Builders tried to set up their own training programs. But this end-run around the unions collapsed precisely because of union-sponsored boycotts. In 1898, the Boston MBA established an Employment Bureau to register tradesmen. Again, this alternative to the union hiring hall failed to attract a critical mass of workers and never got off the ground. Too many contractors were unable to see beyond their next progress payment to make the long-term investment implicit in an employer-controlled hiring setup.[18]

It would be a mistake to overplay the impotence of the nation's contractors. The brick wall thrown up by the Boston MBA in 1886 and 1890 overwhelmed lengthy and well-organized strikes. Clearly, the collective strength of united building employers was capable of running roughshod over any nineteenth-century construction union. And like factory owners, builders had taken steps to stem the tide of unionism. In 1887, William Sayward organized the National Association of Builders. In five years, the NAB was a powerful force with a membership of thirty-five hundred firms. But employers who were often competing for the same building contracts found it difficult to sustain organizational unity. The depression also set back organizing among contractors, thinning the ranks of employers' associations as well as unions. In fact, the unions recovered considerably more quickly. Testifying before the U.S. Congress in 1899, P. J. McGuire characterized contemporary employer organizations as still weak.[19]

They regained strength in the new century. Beginning in 1900, builders in New York and Chicago locked out building tradesmen in order to roll back union gains. These counterattacks spawned a national open-shop drive in construction. The first local sign of impending union reversals came in the Fall River strike of 1900. On May 27, three hundred union carpenters struck for an eight-hour day, joining the city's plumbers already on strike. The planned strike day had been pushed forward after Leeming & Jones fired thirty carpenters who had refused to work with a scab plumber. The walkout initially took its normal course. The men found jobs in nearby New Bedford and Taunton, and in Bristol, Newport, and Providence in Rhode Island. Predictably, a few contractors reduced the hours on their sites after the first week. But the rest of the employers failed to fall in line. By the end of June, the strike was lost. Though the men won their jobs back, they still worked nine-hour days. The Fall River strikers were not beaten by an overt open-shop campaign. Instead, a combination of a building slump, lack of expected financial support from the national union, and the disappointing absence of solidarity from masons, bricklayers, and plasterers doomed the carpenters and plumbers to isolation and defeat. The results gave heart to employers yearning for an end to union expansion.[20]

In the early spring of 1904, a group of builders convened in the scenic Connecticut Valley to reinvigorate an empty shell of a masters' organization. They vowed to use that season's bargaining sessions to keep the lid on wages and union power. The militant Holyoke and Springfield labor movements were singled out as prime irritants. The previous summer, one hundred carpenters had successfully halted the mammoth Fiberloid project in Indian Orchard until all the nonunion workers were replaced. In addition, Springfield's radical Central Labor Union was growing in numbers and political influence. And finally, during the winter of

1903–4, committees representing contractors and the Springfield carpenters district council had virtually settled on a 50-cent increase (from $2.75 to $3.25 a day) to go into effect May 1. In April, a delegation from the New England association of builders convinced their Springfield and Holyoke counterparts to retract the upcoming raise. They encouraged the contractors to take the offensive, hoping "to bankrupt the unions" and end the "abject surrender to union dictatorship." Bankrolled by the association, builders from the two cities withdrew their earlier offer and refused to consider any raises.[21]

Seven hundred carpenters from Springfield, Holyoke, and Chicopee laid down their tools in May. As in Fall River, the initial prospects looked promising. Some of the men traveled to jobs in New Hampshire and Connecticut. Others stayed and collected strike benefits. Neighboring locals pulled workers off sites run by Springfield contractors. The Springfield Central Labor Union endorsed the strike, warning employers to jettison dreams of crushing unionism. "The unions are here to stay," the CLU announced, "and the employer that reckons otherwise will learn to his sorrow his mistake."[22]

The politically sophisticated Springfield District Council appealed to the general public. In an open letter, the union compared the 25-cent rise in wages over fifteen years to the masters' 60-cent increase in billing-per-man charges. "Our wages only average $650 per year," the Council computed, "and that is not enough to enable us to properly provide for our families, buy tools, and pay for cartage of tools from job to job." The letter called on the public to "judge who is right" and to support the strikers by hiring union carpenters for building work.[23]

This bold plan to bypass the contractors was crafted by business agent Walter LaFrancis. In the first week of the strike, forty-five union members formed a house-building cooperative. Offering to supply materials and labor, they advertised low-cost, high-quality residential work built at the new rate of $3.25 a day. They promised "to demonstrate how cheaply good houses can be constructed when the profits of several contractors and middlemen have been eliminated." Mindful of the competitive threat, the union assured the MBA the co-op would be dismantled if the strike were favorably resolved.[24]

Bolstered by the national open-shop drive and local and regional business support, the MBA held its ground. When independent contractor Thomas Gifford agreed to pay the new scale, an MBA delegation pressured him to reverse his decision and brought him into the association. Notified of the union's proposed co-op, the contractors convinced area materials suppliers to raise lumber prices for non-MBA members. Though only marginally successful in recruiting nonunion strikebreakers, the heavily subsidized employers were better prepared to survive a long standoff.

The unions scrambled to hold the strike together by late June. Regular meetings of Springfield and Holyoke businessmen in support of the MBA provided a continuous flow of financial support for the open-shop campaign. Carpenters abandoned hopes of a wage increase and focused instead on maintaining the closed shop. After fourteen weeks, the bitterly divided carpenters narrowly voted to return to work. Gleeful employers offered them work at the old rate and under strict open-shop conditions. In a speech to the Board of Trade, F. W. Job, secretary of the Chicopee builders' association, reported on the "happy lot of master carpenters" over the strike's resolution. Job proclaimed a new era in construction in the region based on open hiring policies, that is, no discrimination against nonunion carpenters, no limits on output or the number of apprentices, and no sympathetic strikes. Accustomed to the closed shop, union members seethed at the terms of the settlement. Reports of assaults on scabs surfaced in the press. Many carpenters packed their bags, never to return. Those who remained worked to see that the open shop "would exist in name only."[25]

The lesson of the Springfield-Holyoke-Chicopee conflict was clear. Contractor victories required regional support from builders' associations and local support from non-building employers. A group of local contractors, left to its own devices, was still no match for a well-organized union backed up by a dedicated labor movement. In 1906, Pittsfield carpenters sought to elevate their $2.50 daily rate to the $3 level of all the neighboring Berkshire towns—Great Barrington, Lee, Lenox, Adams, North Adams, and Williamstown. The contractors offered to pay the higher scale to the "best men" but rejected a higher standard wage. Armed with offers of sympathetic action from the Pittsfield Central Labor Union and Building Trades Council, and the Berkshire District Council of Carpenters, the Pittsfield men opened their strike with the usual array of tactics. They took out-of-town jobs or offered to work directly for the public at the $3 rate.

The contractors' empty threat of an open shop had little effect. They had done precious little organizing. Employer unity was so fragile that the unions were able to lure higher-paying Berkshire area builders into Pittsfield to take over construction sites stilled by the strike. The dominoes started to fall after three weeks. Individual contractors unilaterally instituted the union demand. One builder was quoted as saying, "I am about sick of holding out against the union's demand." Within a few days, a verbal agreement was reached, granting carpenters $3 for an eight-hour day. The settlement was a complete triumph for the union. The only concession salvaged by the contractors was an understanding that business agents would notify job foremen before entering a site.[26]

Pockets of open-shop construction remained in Massachusetts through the years leading up to World War I. Builders on Cape Cod, in

Worcester, and a number of other towns outlasted repeated efforts to enforce union work rules. Yet the unions always survived and prospered. Worcester locals 23 and 877 were strong and active organizations. Joint meetings regularly drew six to seven hundred carpenters. Their Carpenters' Hall, with two large meeting rooms, served as the center of activity for the city's labor movement. New locals in the Worcester County towns of Westboro, Southbridge, Webster, and Grafton sprang up in the shadow of the state's major open-shop city. The locals were effective enough to keep wages and hours on a par with closed-shop strongholds. As elsewhere, carefully planned strikes brought desired results. In 1916, for example, three hundred union carpenters in Worcester struck for an 8¼-cent raise. One day later, their wish was granted. For all their organization and militance, however, Worcester's unions could not make the final step. They never wrangled a closed shop from the local Builders Exchange and the city continued to be one of the state's few hospitable environments for nonunion contractors and workers.[27]

The open shop came and went in other areas. Beginning in 1908, Haverhill builders announced that they would no longer "discriminate" between union and nonunion carpenters. In a strike the following year, the builders turned to Boston's Master Builders Association for assistance in maintaining the open shop. The MBA quickly sent a trainload of nonunion men to take the places of the striking unionists. Unfortunately for the Haverhill contractors, the Boston employers had lied to the scabs, telling them the strike was over. When a large crowd of union carpenters met the train at the depot, the strikebreakers willingly accepted the union's offer of return fare. The strike ended in a compromise, but the contractors had played themselves out. By 1913, the earlier demands for wages and hours had been accepted and, according to the Haverhill *Eve-*

ning Gazette, the city's carpenters were "so well organized" that the closed shop was not even an issue.[28]

The union grew rapidly in the prewar years. A state report described an annual growth rate of two thousand union carpenters as "normal." By 1917, total membership in Massachusetts had risen to 18,655.[29] Not every bargaining session brought success; nor did every strike. But wages steadily climbed, hours dropped, and union control over the job site strengthened. Most of the labor conflicts concerned wages and hours, but carpenters also walked off jobs to protest the presence of nonunion workers, in sympathy with grievances of other union construction workers, to claim jurisdictional precedence over other crafts, to force employers to pay the standard wage rate, to refuse to handle materials from mills in the midst of labor disputes, and to protest "pusher" foremen.

Some strikes involved whole cities; others a single building project. Some succeeded, others failed. In the fairly typical year of 1906, 2,171 carpenters in over a dozen cities struck for a total of 169 days. The issues ranged from five hundred Lawrence carpenters seeking to boost their $2.50 a day rate to twenty men on a job in North Attleboro refusing to work with nonunion workers. The Lawrence tradesmen won a $3 scale after twelve days. In North Attleboro, nonunion carpenters filled the positions of the strikers. The success/failure ratio showed no consistent pattern, varying from year to year and town to town. Of the nineteen carpenters' strikes in 1910, six failed. Victories tended to come a little easier in disputes over bread-and-butter issues; conflicts over the right to enforce union rules were tougher to win. In 1910, for example, only two of the twelve strikes over wages and hours failed. On the other hand, in 1909 and 1910, carpenters were only successful in driving nonunion workers off the job in two of five cases. Similarly, four of the five sympathy strikes in those years failed.[30]

Union leaders called virtually all of the walkouts. But strikes became such a matter of course that, on occasion, rank-and-file members simply acted on their own convictions, violating union orders and agreements with contractors. At the beginning of a 1909 strike of Pittsfield bricklayers, masons, and laborers, Carpenters Local 444 decided against sympathetic action. Nonetheless, scores of carpenters chose to leave their jobs when nonunion bricklayers were brought in. In 1916, Fall River carpenters employed by A. H. Leeming walked out after a nonunion foreman was hired despite an oral agreement between the city's contractors and carpenters local accepting an open shop and outlawing strikes.[31]

May 1 was "opening day" for walkouts. The beginning of the building season was the optimum time to strike. Disputes were usually settled in a week or two. Few lasted into September. In many instances, there was no clear ending date. If a strike was lost, carpenters found work elsewhere while replacements filled their old slots. Even in triumph, improved conditions sometimes just glided into place. In May of 1910, Springfield contractors rejected a request for a half day on Saturdays. Rather than striking, carpenters took advantage of temporary labor shortages to challenge the builders without a loss of pay. They vanished as a group from job sites across the city at noon on the first Saturday in May. They returned on Monday morning and, facing no reprisals, went back to work. The same tactic was repeated the following Saturday and Monday. After two weeks, with no words exchanged, union carpenters had unilaterally instituted a forty-four-hour week.[32]

By World War I, Massachusetts carpenters had effectively built their unions. The long strikes of 1900–1909 did not always end in victory, but the cumulative impact of all these conflicts was the clear installation of the United Brotherhood as the permanent representative of the state's carpenters. Union carpenters have continued to wield the strike weapon to this day, but have never matched the frequency and duration of the walkouts in this era in any other period. The unions largely accomplished what they set out to win—higher wages, shorter hours, and above all, the regulation of the labor market through the closed shop.

Single-family house on Boylston Street in Cambridge, 1910. CHC

(*Opposite page, above*) Reconstruction of Clyde Street pier in East Boston, 1907. SPNEA, B&A

(*Opposite page, below*) Cambridge Gas & Light Co. construction crew. CHC

(*Right*) Pile-driving rig in Natick, 1907. SPNEA, B&A

Construction at the Charles River Dam, 1907. CHC

Pumpwell at Kendall Square in Cambridge, 1909. CHC

(*Below*) Bridge-building on the Charles River, 1908. CHC

Excavation in Harvard Square for the Boston Elevated Railway line, 1910. CHC

Work on the elevated line at Brattle Street, 1910. CHC

Planks strewn along Massachusetts Avenue near Central Square, Cambridge, 1910. CHC

Hundreds of wooden piles form foundation of Eliot Square Carhouse in Cambridge, 1911. CHC

6

Inside the Union Hall

In any workplace conflict, employers retain the ultimate power of dispensing or withholding employment. Workers' ability to modulate that power rests, in part, on the scale of the opposition's resources and their own internal organization. In the decentralized world of construction, building tradesmen held a critical edge. Local and regional employers had difficulty marshaling adequate financial and political clout to wage all-out war on building trades unions. The highly skilled nature of the work reinforced the rough parity of the contending sides. Carpenters and other craftsmen relied on the leverage of their indispensable savvy to drag grudging contractors to the bargaining table. From time to time builders dealt their employees severe setbacks, but they were unable to extinguish permanently the flame of unionism.

For all the advantages built into the situation, the successes of the Carpenters Union in its first forty years rested on the complete organization of the trade. The presence of qualified nonunion carpenters always returned the upper hand to the contractors. As long as the people who had the skills to put a structure together were organized, building employers were forced to negotiate with the union. Union control over the labor supply was a necessary but not quite sufficient condition for success. Locals survived and prevailed because carpenters *wanted* to join the UBCJA. They welcomed the benefits, enjoyed the internal life, and considered union membership part and parcel of the identity of an accomplished mechanic. The union card was as crucial to an early-twentieth-century carpenter as a complete set of tools. He needed the card in closed-shop towns to practice his trade. He needed it in *every* town to be considered a tradesman. By the time of World War I, it was a matter of common wisdom that "the craftsman without a card is a man without a trade."

Founder P. J. McGuire's conception of the United Brotherhood combined elements of craft pride, beneficial unionism, and social vision. Locals wrote sick funds and accident and death benefits into their by-laws as soon as they were chartered. These insurance schemes tided members over the humps of a dangerous and unstable working life and added immeasurably to the appeal of the organization. The extent of the funds belied the infancy of the unions. In 1904, Boston's Local 33 distributed $1,800 in sick and accident benefits. McGuire challenged impatient members who criticized the elaborate benefits system. In 1899, Lynn carpenters actually withdrew from the Brotherhood for a period of five years. Local 108 members objected to the 20 cents monthly per capita tax and argued that "too much money was being spent on death and disability benefits and too little for organizing purposes." McGuire disagreed. He stressed the connection between the insurance funds and broader labor concerns of social change. The benefits, he claimed, "hold enough men together to keep the machinery in motion" in slack times, so that when work picked up and the larger "labor question" could be faced directly, "these men will find an organization all in order for them."[1]

Despite his emphasis on beneficial measures, McGuire did not equate trade unions with insurance societies. Labor organizations, he believed, had a grander purpose—they needed to play a role in the development of an American working-class political culture. In speeches and writings, he contrasted the cultural poverty of workers in the United States with the public libraries, art galleries, and labor parties available to the European workingman. "This is a country," he told a U.S. Senate committee, "that is chiefly intent on making money." He urged locals to set up libraries, train members in the art of public speaking, and sponsor debates on the political and economic issues of the day. "What better field for the discussions of these grave problems than the Trade Unions . . . these primary schools of industrial thought?"[2]

Massachusetts carpenters took McGuire's advice to heart. Boston carpenters joined with the Knights of Labor to set up a reading room for labor literature in the Hyde Park neighborhood of Boston in 1887. Social reformers such as Martha Avery regularly addressed the weekly union meetings. The Worcester locals installed a reading room in their Carpenters Hall. Their meetings combined union business, singing, instrumental performances, and educational programs. Greenfield's carpenters held monthly "smoke talks" devoted to political discussions. The Springfield District Council had a standing committee on books and magazines that purchased works of interest for union members.[3]

Debate and discussion occasionally moved into the electoral arena. A number of union carpenters stumped for political office in the 1890s, a period of heightened labor political activism. W. J. Shields, a national UBCJA officer as well as president of Local 33, ran for lieutenant-governor on the populist People's party ticket in 1891. The depression of 1893 made a legislative orientation even more attractive as traditional collective bargaining brought few or no gains. Carpenters and other unionists turned to radical political initiatives, hoping their candidates could speak for working-class interests on the federal, state, and local levels. In early 1894, the Boston Carpenters District Council issued a call to the membership for increased political and economic discussion and pushed the Building Trades Council to endorse political parties outside the Democratic and Republican two-party structure. The BTC also announced its support for the nationalization of gas and electric utilities and the communications industry.[4]

Following the lead of the building trades, a number of Boston area labor organizations met in April of 1894 and formed the Workingmen's Political League. With the backing of the Massachusetts AFL, the League drew up a mildly socialist platform that highlighted the working-class elements of the People's party programs. The WPL nominated Local 33's Harry Lloyd for the Boston School Committee in 1895 and built enough support that Lloyd lost by just 662 votes. As the depression eased, interest in labor candidates dwindled. Carpenters returned to workplace-based strategies to improve their lot. Nonetheless, contrary to the standard image of conservative building trades unions' hostility to any kind of political activity, Massachusetts carpenters never gave up a strong tradition of involvement, or for that matter, developing qualified candidates. In 1909, for example, John Potts of Local 33 followed in the footsteps of Shields and Lloyd, running a strong race for the Boston Common Council.[5]

The locals took up light as well as grave matters. Committees organized oyster suppers, clam bakes, and turkey dinners that drew hundreds of members and their families. Whatever other events a local sponsored, Labor Day was *the* annual occasion for carpenters to parade their collective pride before their fellow unionists and townspeople. Labor Day marches and picnics were giant affairs—thirteen thousand in Boston in 1896; one thousand in Lawrence the same year; ten thousand in Holyoke in 1902; and forty-five hundred in Brockton in 1903. Inevitably, union carpenters formed one of if not the largest contingent in the parade.

Who were these men who marched on Labor Day? Where did the carpenters of Massachusetts come from? Like most American workers, they were the descendants of immigrants. But like most skilled workers, relatively few were immigrants themselves. The census report of 1850 shows that fully two-thirds of Boston's carpenters were born in Massachusetts or other parts of the United States. Seventeen percent came from Ireland, a reflection of the large community in Boston that fled the Irish famines of the 1840s. Another 15 percent were of British North American origin, primarily arrivals from the Maritime Provinces of Canada.[6]

In 1850, the Canadian connection was just a trickle. The collapse of New Brunswick's timber industry in the previous decade had dispatched unemployed loggers and woodcutters southward. Boston's modest building boom in this period provided a welcome haven for men accustomed to working with wood. That was only the beginning. Over the next century the trickle cascaded into a regular pipeline of carpenters from New Brunswick, Nova Scotia, and Newfoundland. The volume ebbed and flowed with the state of the extensive wood products industries in the Atlantic Provinces, but the stream never completely dried. From the days of the shipbuilding slump in the 1870s right up to the present, carpenters have deserted destitute Maritime villages in favor of the construction jobs in the cities and towns of Massachusetts. Ernest Landry grew up in a small settlement near Moncton, New Brunswick. "If they come from Canada," he observes of his fellow travelers, "it seems that they're carpenters."

Unlike European immigrants, the English-speaking Canadians had little trouble with their new homeland's language and culture. Furthermore, the distance was small, allowing regular contact and frequent return visits. Not all the Canadians spoke English, however. A significant number of the immigrants were French speaking. By 1873, an estimated two hundred thousand French Canadians lived in New England, concentrated in a few towns such as the "Canadian centers" of Fall River, Holyoke, and Worcester. The French communities grew rapidly, particularly during the great migration of 1890 to 1893. They came from Quebec and the French areas of the Maritimes. Moncton, Landry's hometown, was mostly French speaking. Whether farmers, loggers, or fishermen, Landry recalls, the people he grew up with seemed to have sawdust in their veins. In the summer, they built and repaired barns, roofed, shingled, and put up additions. During the long winters, they made wooden strawberry and blueberry crates. Landry's impressions are confirmed by statistical surveys. According to a study of French Canadians in Massachusetts, an unusually high 6 percent of all French Canadian immigrants in the commonwealth in 1895 were employed as carpenters and joiners.[7]

The linguistic and cultural differences made social integration much more complex for French Canadians than for their English-speaking countrymen. The Québecois who moved into the shoe and textile mill towns lived in insular neighborhoods retaining their customs and traditions. The ever-present option of a return to the nearby Canadian homestead worked against the kinds of assimilationist tendencies that were slightly more pronounced among non–English-speaking European immigrants. As a result, many native New Englanders viewed French Canadians with enormous suspicion. Their perceptions were reinforced by the common reluctance to exchange Canadian for U.S. citizenship. In addition, factory workers feared the wage-cutting threat of a new and sizable community of potential competitors.

Anti-French nativism received a boost with Carroll Wright's infamous 1881 remark, labeling French-Canadian immigrants "the Chinese of the Eastern States." In his annual report, Wright, who was the chief of the Bureau of Labor Statistics in Massachusetts, suggested, "Their purpose is merely to sojourn a few years as aliens, touching us only at a single point, that of work, and, when they have gathered out of us what will satisfy their ends, to get them away to whence they came, and bestow it there. They are a horde of industrial invaders."[8] Wright's comments hit a highly sensitive nerve. In the ensuing uproar, leaders of the French-Canadian community forced the bureau chief to hold public hearings on the subject and print their counterpoints in the following year's Statistics of Labor report. Wright's views were far from uncommon. Even unionists in mill towns wondered if such a "sordid and low people," as Wright described them, could forgo a cultural clannishness and forge ties with other factory workers. The notion of an "industrial invader" fed into a stereotype of French Canadians as indifferent or hostile to unionism.

The experience of Massachusetts carpenters' locals refutes the antiunion albatross. The Brotherhood organized separate ethnic locals across the country in an effort to reach out to immigrant workers. Springfield Local 96 was not only the first French local in the commonwealth, it was also the first carpenters' local of any kind in the Springfield area when it was chartered in 1885. Springfield's English-speaking carpenters did not organize until eleven years later. Very rarely, tensions arose over the French locals' willingness to accept lower wages than the English-speaking unions. Much more often, the French locals spearheaded some of the most militant labor struggles in the state. By the early 1900s, French locals had been established in Holyoke, Chicopee, Worcester, Fall River, Lowell, Leominster, and Taunton. Ties to the old country both helped and hurt organizing efforts. Carpenters freely drew on prolabor traditions in Quebec, inviting Canadian union leaders and sympathetic politicians, such as Member of Parliament Alphonse Verville, to speak at their meetings. On the other hand, the virtually open borders guaranteed an influx of nonunion carpenters each spring. The French locals constantly pressed the national Brotherhood office and the Massachusetts State Council of Carpenters to hire French-speaking organizers to bring in the unorganized men.

From its inception, the Brotherhood created foreign-language local unions to organize immigrant carpenters. By 1894, 70 of the national union's 597 locals were ethnic unions—48 German, 9 French, and a scattering of Bohemian, Scandinavian, Jewish, Dutch, and Polish.[9] Massachusetts contributed more than just its French locals. In the early 1900s, Portuguese carpenters won a charter in Fall River and Jewish locals functioned in Boston and Chelsea. The by-laws of the Boston local offered members the choice of using English or Yiddish at meetings, but required English for all officials records and correspondence. The ethnic locals buffered the immigrant carpenter's contact with an alien environment. Unfamiliar with the language and daily patterns of his new American compatriots, the immigrant carpenter relied on the cultural solidarity of his local union to help him find jobs and make sense of his new world. Even in towns without separate ethnic locals, large blocs of single nationalities within an English-speaking union performed the same function.

Ellis Blomquist's father was a construction supervisor and member of the carpenters' guild in Finland. Fleeing to avoid conscription into the Russian army in 1914, he brought his family first to Canada and then to the lumber camps of Massachusetts. Through contacts in the Fitchburg Carpenters Union (at the time half-Finnish, according to Tom Phalen), he quickly became a Brotherhood member. "My father got a job with a Worcester firm," remembers Blomquist. "The superintendent liked him so much that he bought a Finnish-American dictionary, and at lunch hour, he was teaching my father English." But even without a friendly supervisor, strong ethnic bonds generally eased the newcomer's path into the industry. "There were a lot of Finnish people," says Blomquist with a chuckle. "So my father could always get along by being paired up with another square-head."

The ethnic locals faded in importance as the members assimilated into the dominant American culture. The children who followed their footsteps into the carpentry trade no longer needed the cultural buffer of a foreign-language union. Most of the ethnic locals maintained their charters while gradually switching to English for union business. A few retained their original flavor but only at the expense of new blood. Boston's Jewish Local 157 kept its ethnic identity into the 1960s. Most of its members, however, were in their sixties. The youngest were in their late forties. Their sons and nephews either joined the more vital regular locals or moved to fill openings in the professional and business world. A 1960 study considered the state of this Boston ethnic local that did not really die, but rather faded away.

Many of these older carpenters have resigned themselves to working only on the less arduous aspects of the trade—e.g. on wooden buildings and alterations. In fact, some of the less sturdy members are not even interested in "concrete" jobs, or anything that requires climbing. Almost one-half of these carpenters do jobbing on their own—mainly residential housing repairs. Often three or four men will form a partnership, and they are satisfied with "as much as they can make."[10]

In 1972, Local 157—little more than a paper organization by then—was consolidated into Local 33.

Immigrants still form an important element within the trade. In Massachusetts, drywall installers are often first- or second-generation French Canadians or (in the southeastern part of the state) Portuguese. Migration patterns, today as before, depend on family and village ties. Though the current number of Irish newcomers in Boston hardly measures up to the great migrations of yesteryear, most carpenters arriving from Ireland find their way to Local 67, the union with the greatest number of Irish members. For most nationalities, the first carpenter to "make it" invariably serves as an inspiration and contact for those who follow. Virtually every carpenter hired by one present-day drywall firm in the Fall River–New Bedford area comes from the same small town in Portugal, a fact attributable to the drawing power of the Portuguese owner. Today's immigrants, however, must depend on community agencies to cushion the culture shock. The internal life of the modern union has little room for the small minority of foreign-language-speaking members. The tradition of ethnic locals in the Brotherhood has long passed.

For native and foreign-born alike, the unions were more than collective bargaining agents. They were the social center of carpenters' lives. Living memories cannot reconstruct events dating back to the beginning of the century, but today's retirees recall a cultural vitality that continued well into the 1920s and 30s. John MacKinnon joined Boston's Local 67 in 1921.

He remembers the big crowds at the Friday night meetings that stayed on into the night after the union business was finished, talking, arguing, and playing cards.

The union used to be a wonderful place to go to a meeting back in the days when I joined. We had some smart men. It was interesting to hear the conversations and the work that was discussed. You got up at the meeting if you had anything to say and you expressed yourself.

Tom Phalen joined the Fitchburg local in 1927. He was struck by the membership's general knowledge.

Everyone was a parliamentarian. You didn't open your mouth without somebody jumping down your throat, saying "you're out of order or you got to do it this way or that way." Everyone knew the by-laws, everyone knew the Robert's Rules of Order.

Phalen points out that the union both educated and entertained the members at a time when workers had few other outlets. "In the twenties and thirties," he says, "the old-timers didn't have television or anything. They were always at the meetings, and they knew what they were talking about. They would talk union day and night."

"Talking union" meant carpenters first, the combined building trades second, and the entire labor movement third. The carpenters were the largest of the building trades unions and the building trades anchored every Central Labor Union in Massachusetts, even in the mill towns. The strength of the turn-of-the-century Massachusetts labor movement was tangible. Workers of differing industries were connected through their organizations and provided crucial support for one another. In his 1901 novel *The Evolution of a Trade Unionist*, Boston labor editor Frank Foster described the excitement of a "cosmopolitan" Central Labor

John MacKinnon and dollhouse he built in 1975

Union gathering. Carpenters, bricklayers, printers, cigar makers, and dozens of other craft workers of all nationalities convened regularly.

There were radicals and conservatives, cranks and common-sense men, the diffident and the assertive, the wit and the bore, the cynical and the enthusiast, the fault-finder and the organizer, the impetuous and the self-contained, men who rarely spoke and those who delighted in the sound of their own voices. . . . There was no question too small or too large to wrestle with. [The CLU] tackled with equal zest the issue of Cuban independence or that of whether a barber painting his own shop infringed upon the jurisdiction of the craftsmen of the painters' calling.[11]

Carpenters raised money for coal handlers, typefounders, shoe cutters, and textile workers. They patronized grocers with union clerks and helped win a 1902 Boston brewery workers' strike by boycotting nonunion beer. They smoked union cigars, shopped at union stores, wore union clothing, and ate at union restaurants. Parents passed on union traditions to sons and daughters. "Nothing came into our house that didn't have a union label," says

Oscar Pratt. Paul Weiner had the same experience. "You'd go into a store to buy a shirt and it had to have a union label. We were brought up that way." Tom Harrington remembers his childhood household as being "strictly union." The traditions traveled from the home to the job site. MacKinnon says of his fellow carpenters, "we always bought union overalls."

A system of rules and regulations kept the job sites "strictly union." Business agents were empowered to "bring charges" against any member who violated union by-laws or trade rules. In late 1906, a rash of nine-hour jobs cropped up in Springfield. The city's Carpenters District Council ordered agent W. J. LaFrancis to crack down on offenders. Within three weeks, LaFrancis identified eight union members who had agreed to work the extra hour. The Council filed charges, held hearings, found all eight guilty, and fined them $15 each. Locals penalized members for piecework, lumping, accepting subunion wages, and working alongside nonunion carpenters. Complaints against members from other unions were considered seriously. When the Springfield Painters Union informed the Carpenters District Council that two UBC members had hired nonunion painters to work on their own property, the Council promptly issued suspended fines. Travelers from outside a union's geographical jurisdiction who failed to buy a 25-cent working button were customarily fined from $2 to $5. The penalties for transgressions could be severe. In 1908, LaFrancis discovered that member C. Junior had accepted a nine-hour open-shop job. For the double crime of working alongside nonunion men and for nine hours a day, Junior was ordered to pay $30—nearly two weeks earnings.[12]

As membership grew, union business agents were hard pressed to patrol every job in their areas. A 1907 Springfield District Council by-law gave the business agent authority to appoint job stewards as the on-site eyes of the agent. The rules were later altered so that the first journeyman on a job acted as steward. Once three carpenters were hired, they elected one to serve for the duration of the project. The Worcester system was a slight variation on the same theme. The first carpenter automatically became steward. In case a group of men arrived together at the beginning of construction, the one with seniority of union membership took the post. They were working carpenters but they had additional union-related responsibilities. The steward checked dues books to ensure payments were up to date. In addition, he reported any infraction of union rules. Though he lacked the power to call walkouts unilaterally, the steward's observations determined many of the business agent's actions. Inevitably, aggressive stewards ran the risk of alienating their employers. In order to protect job security and ensure militance, local by-laws often required that "union men must stand by the steward in the performance of his duties."[13]

This system of justice and enforcement served as a code of ethics for union carpenters. Sharing a job site with nonunion men was wrong, pure and simple. From a strictly pragmatic perspective, nonunion workers threatened union members' security and wage levels. But more than that, men who never entered the union halls stood outside the community of union values. Similarly, members who chose to be pace-setters, pushers, or pieceworkers rejected, by their very actions, the culture of solidarity and cooperation that supported the framework of unionism. The system's charges, hearings, and fines had two purposes. They punished the guilty and symbolically reminded the membership of their shared principles. For those reasons, the process was taken seriously. Hearings on charges were formal in tone and afforded violators the opportunity to defend themselves. Few contested the accuracy of the charges, and even fewer contested the underlying basis of the rules. Fines were paid and life went on. But as with every system of justice, allowances were made and rules were bent if necessary. Sometimes the exceptions were even written into the laws, as in the issuance of privilege cards to "aged and crippled brothers" who were exempted from the ordinarily absolute rules governing minimum standard wages.[14]

The union culture extended to the broader labor movement. An active labor press articulated the labor creed and kept union members informed of current events. *The Labor News,* begun in 1906 by union typographers on strike at the *Worcester Telegram,* was widely read until its demise in 1972. Frank Foster started Boston's *Labor Leader* in 1887, somewhat before the Lynn Central Labor Union issued the *Lynn Union Leader.* The *Labor Advocate* appeared in Springfield from 1917–1920. Beyond the general labor press, the Springfield and Greenfield Carpenters Union locals subscribed to the craft-oriented *Artisan* for their members.

The press covered local union news and national affairs of interest to the labor movement. At its best, the information cemented the ties that bound workers in different unions together. These bonds were particularly significant for construction workers, men who were drawn together by similar levels of skill and common work sites but separated by different employers and labor organizations. Their relationships had always been tense and intense. Workers of the varying building trades walked a tightrope with one another, dancing from selfless cooperation to extreme competition.

The pinnacle of cooperation was, of course, the sympathy strike. In these strikes, one set of tradesmen respected the other's protest, walking off a job site even if they had no grievance with their own employer. Sympathetic actions usually forced an employer's hand more quickly than isolated strikes by individual crafts. The disappearance of any one trade from a project slowed progress gradually. A combined walkout of a number of trades brought construction to a complete halt. In an 1898 study,

Fred Hall discovered that sympathetic strikes occurred more often in building than in any other industry.[15] Common social backgrounds, skill levels, and work sites strengthened cross-trade unity in construction. As a result, building contractors could rarely use the kind of divisive antilabor strategies common in other settings, such as harping on skill, nationality, or race differences. On the contrary, cultural similarities made the obviously effective sympathy strike easier for building trades workers to carry out.

Builders detested the sympathy strike. They wanted, above all, uninterrupted production. Resolving conflicts with their own employees was difficult enough without having to pay the price for someone else's dispute. The threat of sympathetic action pushed the Boston MBA to back off its harsh antiunionism and institute trade-by-trade arbitration committees in 1893. Apparently, the issue was sufficiently important to warrant major compromise. MBA spokesmen recognized that organizing administrative bodies by craft rather than cross-industry substantially weakened employer leverage. However, they considered the pledges to outlaw sympathy strikes worth the concession. The 1902 agreement between the United Carpenters Council and the Master Carpenters continued the ban on strikes, requiring all disagreements to be resolved by arbitration.

Nonetheless, Boston contractors never completely wiped out the sympathy strike. In June 1907, a group of carpenters won a two-week strike after they refused to work with materials from a mill that was on strike. One month later, fifty carpenters, masons, bricklayers, and hod carriers convinced a subcontractor to replace his nonunion workers after a five-day walkout. In 1912, 252 carpenters, plasterers, bricklayers, metal lathers, and hod carriers supported a steamfitters' jurisdictional claim in an eight-day walkout; twenty-seven carpenters staged a one-day strike to protest nonunion bricklayers; and eleven carpenters, plumbers, and electricians left their jobs rather than work with nonunion painters. In all three cases, the supportive actions led to the correction of the problems.[16]

Outside of Boston, sympathy strikes were less successful though just as frequent. The Massachusetts Bureau of Statistics of Labor reported six sympathy strikes involving carpenters outside the capital city between 1909 and 1912. There was no particular pattern. The towns and the projects varied in size. The trades under siege included bricklayers, electricians, painters, plumbers, and laborers. Each walkout failed and the strikers found jobs elsewhere. Yet the principle remained intact. If one trade was threatened—particularly by nonunion workers—other trades provided support on request. This was as true in open-shop Worcester as in closed-shop Boston. In 1914 and again in 1915, Worcester carpenters struck major building sites over the presence of electricians employed by nonunion Coughlin Electric.[17]

Sometimes, the very possibility of a sympathetic strike carried as much weight as the action itself. For a period, carpenters locals in various parts of the country respected agreements with bricklayers, masons, and plasterers under which no one trade could work if any of the others had a current grievance with an employer. The pacts were respected in Massachusetts. The Springfield locals went one step further—the understanding extended to the Building Trades Council and the Central Labor Union. Thus, if any one of the four trades chose to strike, the entire organized workforce of the city would walk out in sympathy. The details of the pact were public knowledge. Many observers credited the successful building trades negotiations of 1914 to the widespread contractor fear of precipitating a general strike.[18]

Relations between the craft unions had a nastier, warring side. The sympathetic shows of solidarity may have enraged builders, but the constant jurisdictional fights between trades baffled and infuriated the general public. The sheer volume of construction activity silenced by workers locked in mortal combat over seemingly trivial work assignments has prompted more antiunion sentiment over the years than any other aspect of the industry. Battles have raged between craft unions, between building and industrial unions, and between locals of the same union. Trades regularly crossed each other's picket lines or refused to work alongside craftsmen from another union if a jurisdictional battle had not been resolved. The ludicrous sight of fellow unionists bringing employers to their knees over internal conflicts led members of the Industrial Workers of the World and other labor radicals to label building trades unionists "union scabs." No other struggle in the field has produced the passion and rancor of these jurisdictional disputes.

The vast majority of jurisdictional conflicts stemmed from competing claims over new materials, new machinery, or changing construction methods. With every craft in a state of perpetual flux and every craftsman fearing for his future, each innovation represented a potential minefield of lost job opportunities. As the twentieth century ushered in a new wave of technological changes in construction, the jurisdictional battlefront heated up. Bricklayers fought with stonemasons over installation of terra cotta blocks, plumbers and steamfitters argued over which pipes belonged to which trade, and carpenters fought with everyone.

UBCJA officials portrayed themselves as besieged on every front. The trend of metal replacing wood created conflicts with the Ironworkers and the Sheet Metal Workers and the introduction of concrete prompted scraps with the laborers. But more than job security was at stake. The rapidly growing number of jurisdictional strikes owed as much to the prevailing ideological winds inside the unions as it did to a commitment to job protection. Under

McGuire, the national policies of the Carpenters Union generally discouraged competition between trades. McGuire was not reluctant to take on other unions, though he usually focused on those organizing carpenters, such as the Amalgamated Society of Carpenters and Joiners, the United Order of Carpenters, and the Knights of Labor. McGuire reasoned that carpenters would achieve optimal power in a single organization and he relentlessly pursued this objective. He did not live to see it reached, but all the other organizations of carpenters had either faded away or joined the United Brotherhood by World War I.

However, McGuire condemned jurisdictional quarrels with other AFL unions. He advocated the general organization of the workforce and the growth of the trade union movement above and beyond the gains of a particular AFL affiliate. As the ranks of the UBC swelled in the 1890s, this catholic perspective put him in conflict with a growing number of other union officers whose sole concern was the fate of the Brotherhood. The most pressing controversy was with the Amalgamated Wood Workers, an AFL union chartered to organize mill hands in the country's woodworking factories. Though the UBC had nominal jurisdiction over mill workers, the union had given them half-hearted attention. At first, McGuire resented the AWW's muscling in on Brotherhood turf. He soon dropped his opposition, recognizing the industrial union's superior organizing efforts.

Others were less charitable. Pressure mounted on McGuire to renounce his fraternal attitude through the second half of the 1890s. When he refused to reconsider, a struggle for succession began. Though differences over jurisdictional policies were not the only reasons for McGuire's ouster from office in 1901, William Huber and Frank Duffy, the Brotherhood's new leaders, gave quick evidence of changed official attitudes at the annual AFL convention later that year. Delegates from the Brotherhood attacked the Electricians, Ship Carpenters, and Painters and Paperhangers, as well as their long-time foes, the Amalgamated Wood Workers and the Amalgamated Society of Carpenters and Joiners. The national office unleashed pent-up hostilities at the local level. Free to pursue empire-building strategies, UBC unions across the country launched a series of jurisdictional strikes and raids. The national union laid claim to a variety of construction tasks that had never before been considered part of the carpenter's job. In 1886, the UBCJA constitution described a potential member as a carpenter, joiner, stair builder, cabinetmaker, millwright, or factory woodworker; in 1919, the equivalent jurisdictional definition took five thousand words. Evaluating the impact of the Duffy–Huber regime, Charles F. Reilly, head of the United Order of Box Makers and Sawyers of America, sarcastically commented: "It's a wonder that they do not claim the earth."[19]

Naturally, aggressive jurisdictional schemes in any one union led to parallel plans of action in other unions. Jurisdictional wars became as familiar a part of twentieth-century construction sites as overalls and tool chests. Conflicts between trades over work assignments were so routine that the secretary of the Boston Building Trades Council wondered if some of the claims should even be given serious consideration. "The general practice of the national union," he observed, "is to claim all in sight for safety's sake."[20] Building employers chafed at their lack of influence over the feuding unions. Occasionally, contractors hinted at a willingness to accept the vastly increased power of an amalgamated construction union of all trades in order to eliminate jurisdictional disputes.

In the blueprints for the State Mutual Building in Boston, the architect specified metal door bucks (frames) throughout the structure. At the time (1914), carpenters and ironworkers were contesting installation of the recently designed metal bucks. When the contractor, Norcross Brothers, awarded the work to the ironworkers, the carpenters on the site walked off in protest. Within days the other four hundred tradesmen ran out of work. Norcross informed both unions it would gladly abide by any decision they made. But in this case, employer flexibility meant nothing. The locals were unable to come to an agreement. Representatives from the respective Internationals arrived and hammered out a proposal in which a composite crew of seven carpenters and seven ironworkers handled the bucks. But jurisdictional wars were rarely that simple; Boston Carpenters' officials rejected the compromise.

In the meantime, Norcross was frantic. The initial walkout occurred on February 20 and no solution was in sight a month later. On March 20, state arbitrator Charles Wood wrote in frustration to AFL head Samuel Gompers:

The contractor is between two fires. He is unable to complete the building under construction for the reason that if he gives the work to the ironworkers the carpenters say they will go on strike, taking with them such trades as are allied to them, and if the carpenters erect the bucks the ironworkers will go on strike, carrying with them the trades affiliated by alliance. If the work is performed by non-union men, a strike would probably follow by all trades.

Finally on March 25, the carpenters agreed to the composite crew. Three days later, C. L. Covington of Norcross advised the arbitrator that nearly all the bucks had been erected.[21]

The jurisdictional wars in construction drew scathing denunciations from contractors, the general public, and other trade unionists. The conflicts grew out of one aspect of building trades craft unionism—the narrow focus on the exclusive interests of the particular trade's membership regardless of the consequences. But contrary to popular perceptions of the Carpenters and other construction unions, the predominant expression of pre–World War I building trades craft unionism on the local level was not the desire for a separate room in the House of Labor, but rather extensive demonstrations of solidarity. The ever-present sympathy strike, the leadership roles in Central Labor Unions, the pride in the statewide Labor Day celebrations, and the regular presentation of cultural and educational events for the entire labor movement are testimony to the culture of cooperation that characterized carpenters' unions in Massachusetts.

But a more useful way of understanding these actions is not whether some of them were "good" or "bad," but that all of them were consistent with the carpenters' choice for a union culture. The union raised the image of the carpenter to a lofty status, as a proud, "manly," independent, and even special worker with a right to be treated with dignity and respect. To the extent that that ethic coincided with the interests of the larger labor movement and the working class a whole, carpenters were exceedingly generous in their gestures of cooperation and solidarity. If, on the other hand, someone threatened to interfere with their explicitly demarcated sense of rights and duties—be it an employer, another building trades union, or a radical industrial unionist—the carpenters' organizations could be unremittingly hostile and vindictive. Individualism and cooperation marched side by side comfortably within the carpenters' union halls.

7

Birth of the Business Agent

In the summer of 1883, an unusual notice appeared in the New York City daily newspapers. The text informed the city's contractors that, from that day on, they were to hire union workers exclusively. In accordance with the notice, a contractor named Corr was pressed to dismiss two nonunion derrickmen on a midtown Manhattan project. Corr refused and two hundred craftsmen walked off the site. After two days of intensive negotiations, the workers returned to Corr's revamped all-union job. The walkout was successful, but, more significant, the action marked the first time that union building tradesmen conducted a strike through the offices of a full-time union representative—the walking delegate.[1]

James Lynch, an officer in the United Order of Carpenters, was the union spokesman in the Corr dispute. He later wrote that his appointment as walking delegate was a move made "in desperation." Lumpers and piecework contractors had invaded New York City bringing with them scores of untrained nonunion carpenters. The unions had tried unsuccessfully to organize the new arrivals by traditional haphazard methods. Discouraged by their failure, leaders of the Amalgamated Building Trades Union

created a new and permanent union office devoted to actions on behalf of the city's organized tradesmen.[2]

The need for such a position had long been apparent. Militant craftsmen who voiced grievances against building employers were often sent packing. Discrimination against early union activists was common. In 1881, a New Haven, Connecticut, carpenter reported that local contractors shared a blacklist not only of union activists but of any workers who left an employer in order to seek higher wages elsewhere. Many of the young UBCJA locals dug into members' pockets to help out brothers who had been fired for organizing. Extensive blacklists, however, kept the numbers in need on the rise and made that solution an expensive demonstration of solidarity.[3]

If only to preserve local unions' treasuries, another approach had to be found. The New York experiment offered a direction. Instead of draining the members' resources for a potentially endless list of martyrs, the New York unions had opted to pay one regular salary. These funds supported the walking delegate (later known as the business agent), a man whose livelihood depended on the union, not the employer. In theory, therefore, his immediate interests lay as an advocate for the membership. Unlike the vulnerable rank-and-file member, he could confront employers forcefully and vigorously without jeopardizing his own weekly pay. Initial success in New York won attention elsewhere. P. J. McGuire told the 1888 national UBC convention that walking delegates were operating in fourteen cities.[4]

In October 1887, Boston's Local 33 elected Chelsea-born Joseph Clinkard as the state's first walking delegate. In July of 1889, two other Boston locals joined Local 33 in selecting Clinkard. Over the next few years, until his untimely death in 1894 at the age of forty-one, Clinkard wore several different hats as the principal spokesman for the city's carpenters. He served as district organizer, general agent for the Boston District Council of Carpenters, and president of the Boston Building Trades Council. Other Boston walking delegates, including C. E. Jordan, J. W. Comstock, and George LaSeur, carried out their duties under Clinkard's protective wing.[5]

The walking delegate was the chief executive officer of the local or district council. He managed the finances, balanced the books, and reminded members to pay their dues. He represented the union in public forums or in private negotiating sessions with employers. He brought up members on charges and called and led strikes. His primary task, however, was to patrol his jurisdiction through weekly visits to the members on the job. At the site, he inspected working conditions, hunted for nonunion workers, determined if union rules were enforced, and listened to workers' grievances. If everything was in order, he continued his tour to the next project. In cases of unfair employers or violations of trade rules, the walking delegate took the heat off individual carpenters and argued their case directly to the contractor.

Walking delegates and business agents also functioned as the chief organizers for their locals. Whether they signed up new members or won dismissals of nonunion workers, they were mandated, as Quincy Local 762 agent John Cogill put it, to get "every man that works with carpenters' tools in the United Brotherhood." Like most other members of the Brotherhood, Cogill understood the value of achieving complete organization. "Not until then," he stated, "will we be able to call ourselves secure." The most active locals insisted that each member act as an organizer, encouraging friends and relatives in the trade to join the union fold. But ultimately, the agent was held responsible for accomplishments and setbacks. The best agents, recognizing the mobility of the workforce, knew their work was not limited to their locals' geographic boundaries. In a letter to the *Carpenter,* Cogill noted that his local's firm control of Quincy's carpenters had prompted him to organize another local north of the city to prevent outside nonunion carpenters from threatening Local 762's position. After a winter of nightly home visits to unorganized carpenters, Cogill had convinced thirty men to apply for a charter from the national office and confidently predicted that another forty members would join the new local within four months.[6]

Before the unions established hard-and-fast labor practices in the industry, employer abuses were multiple in type and widespread in practice. Builders who verbally agreed to respect a standard wage scale commonly undercut the rate by privately demanding kickbacks or distorting the total numbers of hours worked by individual craftsmen. To counter these and other employer maneuvers, walking delegates paid surprise payday visits to witness the opening of the pay envelope. Unaccustomed to intervention by outsiders, contractors rarely welcomed the union representative. In the first years of the new office, walking delegates were harassed legally and physically. In 1892, contractor James Emery literally threw Cambridge Local 183 agent Maloney off an Emery project. Though the local eventually won a minor lawsuit against the contractor, considerably more than Maloney's pride had been hurt in the incident. The job description, therefore, called for a particular kind of individual. As Luke Grant, author and one-time member of the Chicago Building Trades Council, suggested, the "attitude of employers not only made the walking delegate a necessity, but in a degree determined his qualification. He had to be a man possessed of physical courage, who would not be intimidated."[7]

Hostility to the walking delegate was not initially restricted to employers. Nonunion workers feared his presence and some union members resented his role. A conscientious business agent who insisted on work-rule enforcement was liable to embarrass union men who had curried favor with the boss by ignoring union regulations. And inevitably, accusations of selfish ambition and easy living accompanied a union representative's abandonment of the tools. James Lynch evaluated his job in largely negative, if somewhat self-serving, terms:

I found the position of walking delegate anything but a pleasant task. Although naturally of a peaceable disposition I was plunged into a continual war. My presence on a job was an irritation to the employer as well as the nonunion men, and not infrequently some of the union men envied me, not realizing the sorrows of my lot. I retired, after serving four terms.[8]

Within a few years, Maloney's tussle and Lynch's sorrow were small potatoes in comparison with the abuse poured on the heads of business agents. The acceptance of unionism in the industry implied the institutionalization of the walking delegate's office and power. The agent became the lightning rod for antilabor sentiment from every corner. Frank Foster of the *Labor Leader* complained that the new union official had been publicly convicted as "the

promoter and instigator of labor troubles, the ever-present thorn in the side of industry, the arbitrary dictator who rules with merciless power the poor wights under him, the blatant demagogue who is responsible for strikes, lockouts, riots."[9]

Despite the criticism, the office of business agent was often a plum. Like other elective or appointed posts in the labor movement, the position offered an alluring path out of the daily trials and tribulations of the laboring life. Big-city business agents called the shots for a battalion of tradesmen and spent their days negotiating, conversing, and hobnobbing with people considerably higher up the social ladder. The appeal of an office that enabled ambitious working-class men to leave behind the hazards of manual trades should not be underestimated. The fleeting appointments as foremen and/or superintendents represented the outside limits of upward mobility for most carpenters, so leadership of a trade union was a rare prize to be seized and cherished.

The post occasionally offered other gains besides improved status and self-worth. Larger contractors cut deals with business agents in order to consolidate and improve their position in the market. Sweetheart agreements, i.e., the promise by union tradesmen to work only for employer association contractors in exchange for a closed shop, squeezed out nonassociation builders and added to the agents' leverage. Contractors courted the union officials, sometimes sweetening a cooperative relationship with individual rewards. Walter Ohlschager, a Chicago architect, claimed that he always incorporated a 1 percent surcharge for union graft when he estimated building costs. Though Ohlschager's rule of thumb represented a token investment for builders or owners, such a sum looked positively princely to men striving to escape working-class poverty. Luke Grant insisted that most walking delegates were decent, honest, and effective union representatives. But, he admitted, the opportunities in major cities could be heady. The newly elected agent, he wrote, "finds himself suddenly transformed from a position of servitude to one of authority. He has a small army of men at his command. With an exaggerated opinion of his own importance, he is apt to abuse his power before he realizes his responsibility."[10]

Legendary characters like the Ironworkers' Sam Parks in New York and the Steamfitters' "Skinny" Madden in Chicago contributed to the popular portrait of the corrupt walking delegate. Their audacity and unscrupulous tactics brought them considerable wealth and undisputed power within their unions. Parks had a personal "entertainment committee," which once boasted of ninety assaults on dissenting union members in a two-month period. But his long term in office actually rested more on his ability to pressure employers. The Ironworkers won record wage increases during his reign. As one ironworker put it, "Sam Parks is good hearted, all right. If he takes graft, he spends it with the boys."[11]

Hostile journalists and public figures made few distinctions between the effective use of a powerful union office and outright corruption and demagoguery. A sufficient number of early business agents did indeed abuse their powers, but their ranks were largely limited to a handful of big cities where the sheer volume of construction justified "payments for peace" in an employer's calculations. Nonetheless, the Parkses and the Maddens of the building trades titillated the public's imagination and made for exciting newspaper copy.

In 1894, Rudyard Kipling published the widely read "The Walking Delegate" in *The Century* magazine. Kipling's short story crystallized the elements that formed the negative image of the new building trades union officer. In his allegorical tale of the animal kingdom, an interloping yellow horse disrupts the pastoral idyll of a group of Vermont horses. Spouting nonsense about "Man the Oppressor," the outsider rails against "invijjus distinctions o' track

an' pedigree" as barriers to "the inalienable rights o' my unfettered horsehood." The other horses ignore his rantings and efforts to divide the group. But when the yellow horse proves to be lazy as well as foolish, the hard-working and conservative Vermonters "keep school" with him and pound a lesson into his hide.[12] Kipling's trotting delegate is coarse, divisive, slothful, and indifferent to the basic American pride in individualistic values.

Not all of the labor-related fiction of the era was so blatantly antiunion. But even the sympathetic literature reflected popular concern with graft in the building trades. Leroy Scott's novel *The Walking Delegate,* published in 1905, is a sensitive, if melodramatic, presentation of life in the New York City Ironworkers' Union. Scott zeroes in on Buck Foley, the three-thousand-member local's walking delegate. Foley's character is clearly modeled on Sam Parks, yet Scott rejects the superficial stereotypes that Kipling and others favored. In Scott's portrait, Foley is tough and perceptive. He used his native street smarts to rise from the ranks, winning sizable wage packages from employers based on a sophisticated understanding of the growing importance of steel skyscrapers. "Until his union had put up the steel frames the contractors could do nothing—the other workmen could do nothing. A strongly organized union holding this power—there was no limit to the concessions it might demand and secure."[13] The very lack of those limits proved to be Foley's undoing. After four years of exemplary leadership, he began accepting bribes. Soon his lust for money outstripped all other concerns. He organized a clique of thugs to intimidate opponents and ensure his continued stay in power.

Scott's book reflected a need felt by friends of labor to condemn the presence of trade union corruption. Foley becomes, in effect, Sam Parks. Yet Scott managed to be understanding as well as critical of Foley's pact with the devil. His story is one of an ambitious and tenacious worker, led astray by visions of grandeur. The plot of the book and the other characters demonstrate Scott's underlying sympathies with the working ironworker. Tom Keating, hopelessly brave, principled, intelligent, and talented with the tools, is the hero. Initially a Foley supporter, Keating challenged the walking delegate for union leadership, precipitating a torrent of dirty tricks and violence. The story's villain, though, is not Foley, but rather James Baxter, the wealthy socialite and president of the Iron Employers' Association. Baxter, a man who stops at nothing, enticed Foley to sell out the big strike. Scott's condemnation rests squarely on the shoulders of the oily, suave, and manipulative contractor who has none of the excuses of derailed working-class aspirations to explain his evil actions away.

The flood of damning publicity forced the rest of the labor movement into a difficult position. Mindful of the gains directly attributable to the office of walking delegate, labor spokesmen outside the building trades were reluctant to join the critical chorus. The acknowledged presence of corruption, however, diminished the labor movement's ability to garner broad-based support through general appeals for social justice for American workers. Some unions looked the other way. The *United Mine Workers Journal,* for example, compared the righteousness of walking delegates to "Peter and Paul in their sacred work." But few unions adhered to the pure-as-driven-snow line. Most labor leaders fiercely defended the bulk of walking delegates while admitting the reality of abuse. The *Street Railway Employees' Gazette* recognized occasional "over-officious" behavior but firmly declared that the creation of the office was "evidence of progress in the trade unions." Nonetheless, the mounting damage was severe enough to warrant a resolution at the 1902 AFL convention calling for an investigation of union graft. Though the proposal was defeated, a number of delegates felt such an action was necessary in order to refocus public attention on the central issue of workers' rights.[14]

The misguided highlighting of graft among walking delegates was not just a case of over-emphasizing a few bad apples who spoiled an otherwise admirable barrel. Corrupt business agents operated in New York and Chicago because those were among the few cities sizable enough to afford illicit opportunities. Generous payoffs made sense only in situations of massive construction efforts. Even in cities like Boston, business agents never carried the weight of a Sam Parks. Whether it was a matter of personal integrity or lack of opportunities, formal accusations of corruption never surfaced against carpenters' walking delegates in Massachusetts. While men like Joseph Clinkard and John Potts of Boston or W. J. LaFrancis of Springfield were influential citizens in their communities, sharing podiums with politicians and civic leaders and wielding significant power, they did not fit the Kiplingesque image of the "labor boss," ready to shut down an entire metropolis on a moment's whim.

Boston's walking delegates may not have been charged with graft, but that hardly exempted them from merciless criticism. In the midst of an unresolved dispute in 1889 at Boston's Tremont Theater site, Joseph Clinkard ordered a walkout of the project's carpenters. Clinkard had acted to force compliance with union work rules, but a columnist for the *Boston Record* viewed the matter differently. Scornfully deriding Clinkard as "The Walking Delegate, in a plug hat and ponderous watch chain," the writer accused him of arrogant power-hungry behavior and, worst of all, the sin of interfering with a contractor's right to conduct his business as he saw fit.[15]

Perhaps the most celebrated controversy whirling around the city's business agents was the one engineered by Harvard University President Charles W. Eliot. In the midst of the open shop drive of 1903–4, Eliot delivered a stinging attack on unionism in a well-publicized address at the Pauch Gallery Mansion in Brooklyn. The Harvard president charged that "labor unions have wrought more disaster on our country than the Civil War." He singled out the building trades and their walking delegates as wreaking particularly disastrous havoc.

The building trades unions in Boston demand that they decide who shall get certain jobs; that the specifications be submitted for their approval first; that the unions get a certain amount for the awarding of work to certain contractors. Then the walking delegate gets his piece. And if at any time a new combination springs up in their respective unions, a like amount of money must be paid to them as to the old clique; and if such is not forthcoming work is stopped and the contractor is left with the building unfinished until he accedes to the union's demands.[16]

Needless to say, Eliot's words provoked an uproar. The Boston Building Trades Council and the Carpenters' District Council called special meetings to draft a response. Local 33 Business Agent John Potts dismissed Eliot's remarks as ignorant misstatements and challenged him to prove his allegations.

I don't know of a single business agent in Boston who ever got as much as 10 cents from an employer. . . . The employers and business agents in this city don't do business that way. . . . If President Eliot knows of any particular cases here in Boston where the walking delegates have been dishonest, why don't he name the cases and make the matter public, so that the men can be brought into the courts, if necessary. He should not make a sweeping statement as he has, casting a slur on every walking delegate or business agent in the city.[17]

Pressured from all sides to clarify his comments, Eliot agreed to defend himself in a spe-

cial meeting on February 7, 1904. Over two thousand angry unionists converged on Faneuil Hall to attend what the *Boston Globe* labeled a "Trial by Labor." Under heated questioning, Eliot failed to substantiate charges of illegal conduct but refused to revise his basic antiunion ideology. The patrician president's commitment to a laissez-faire liberalism was unswerving as he attacked both trade unions and employer associations as restrictive and "exhibitions of class selfishness." His comments failed to placate the audience. On the contrary, when he termed scabs "a fair type of hero," the Harvard president ensured that the controversy would not die.[18]

As long as distinguished figures like Charles Eliot continued to demean the walking delegate, the taint of demagoguery and corruption covered everyone who occupied the office. For those business agents who labored outside major urban centers, the characterization of raw unbridled power must have bordered on the ludicrous. Most locals could not even afford the luxury of a paid full-time official. Union officers worked in the field suffering the same scrapes, injuries, and falls as every other member and conducted union business in the evenings or on weekends. Old minutes of the meetings of Greenfield Local 549—a good-sized and well-organized local—indicate the enormous chasm between the temptations of big-city union office and the hassles of the small-town equivalent. Until 1910, Local 549 functioned without a walking delegate. That year the local established the post on a one-half day every-other-week basis. The rest of the time, the new officer worked at his trade. Since only a fraction of his duties could be performed in the time allotted, the position turned into a twenty-four-hour job, tucked into any available free moment. The first three candidates, Brothers Barton, Dwyer, and Parnell, each burned out after single terms of one year and "positively refused" to serve again. Parnell cited lack of time as the reason for his resignation. Though members appreciated the walking delegate's contributions, the rewards were minimal. An April 1913 meeting voted Parnell a box of cigars. Finally, after three consecutive resignations, no one could be recruited for the post. In October 1914, the local empowered the union president to appoint (against their will, if need be) any member to the position of business agent.[19]

The Sam Parkses and "Skinny" Maddens grabbed headlines and confirmed antilabor convictions. But their thirst for wealth and power obscured the real significance of a new stratum of union officialdom. Even in the big cities, the typical business agent was not a criminal as much as he was the trade union movement's equivalent of a political ward boss. He mediated between employer and employee, settling grievances and regulating labor relations on the job. He knew the members and he knew the contractors. As the only person with a direct pipeline to both sides, his access to and control of information gave him a prominence unmatched by any other figure in the industry. And when the union had established some form of hiring hall, he distributed the building trades unions' version of political patronage, i.e., construction jobs. In the words of Robert Hoxie:

The peculiar duties of the walking delegate are such as to give him easy ascendancy over the rank and file. He looks out for employment for them; his duties lead him over the whole local field of labor, he knows where jobs are and how to get them, he can keep a man at employment, or he can keep him from it. . . . Clearly he is a man to keep on the right side of, and to keep "in" with. . . . He is a specialist in labor politics, with favors to give and to withhold.[20]

8

Battling Carpenters: World War I and the 1919 Strike

The Brotherhood had demonstrated convincingly that it could organize carpenters and negotiate agreements with contractors in the private sector. The entry of the United States into the First World War in April 1917 posed a new and more complicated set of problems for the trade union movement. The federal government emerged as a primary construction purchaser, redirecting the building industry's priorities from conventional construction to temporary military housing, shipbuilding, and ammunition factories. Bargaining with an individual employer or an association was one thing; taking on the government in the midst of a wartime emergency was another matter.

Building tradesmen were concerned that the new prominence of federal agencies in the industry might undermine established working conditions. The volume of military construction was substantial enough that federal guidelines and standards inevitably set the pattern for all construction. On the eve of the war, Samuel Gompers promised President Woodrow Wilson organized labor's support and, in effect, offered a wartime no-strike pledge. The unqualified character of Gompers's statements concerned UBCJA officials anxious to install safeguards in any wartime agreements. An editorial in the *Carpenter* reminded Gompers that sacrifices for the war effort did not necessarily imply diminishing worker protections. "Patriotic manifestos, unsupported by definite administrative plans, offer no such guarantees," warned editor Frank Duffy.[1]

The Brotherhood did have such plans. The union offered its thirty-five years of experience as a recruiter, trainer, and supplier of labor as a model for an effective and fair wartime labor policy. The May 1917 editorial suggested the unions could play a similar role for federal construction needs: "The AFL, with its great army of skilled mechanics, is in a splendid condition today to be of valuable service if the Government will have the forethought and vision to avail itself of its assistance in a spirit of cooperation compatible with the democratic ideals for which the labor movement stands."[2] From the union's point of view, cooperation entailed a recognition of existing labor conditions, particularly the closed shop.

In June, Gompers huddled with Secretary of War Newton Baker. The results of their conference confirmed the apprehensions at UBCJA headquarters. Gompers had won a federal commitment to prevailing wages, hours, and many working conditions, but specifically exempted the closed shop from the list of prevailing conditions. In essence, the head of the AFL had exchanged a universal union pay scale for open shop operations. Antilabor federal authorities seized the opportunity and issued contracts equally to both union and nonunion contractors in what had been strict closed shop areas. The building trades wing of the AFL was furious with the terms of the Baker–Gompers agreement and none more so than President William Hutcheson of the UBCJA. With a pen stroke, Gompers had wiped out the fruits of years of difficult battles. Hutcheson balked at waiving the closed shop. In protest, he refused to participate on any of the tripartite boards set up to oversee federal labor contracts. A showdown was unavoidable. Sooner or later, union craftsmen would challenge the federal policy.

In early November, a large group of union tradesmen walked off the $9-million open-shop destroyer plant project in Squantum. On November 7, the Building Trades Councils of Greater Boston and Quincy endorsed the walkout and called a general strike on all military work in the area. Thirteen hundred men struck the Watertown Arsenal, the Charlestown Navy Yard, the Federal Appraiser's stores in Boston, the Magazine Station in Hingham, and the Marine Hospital in Chelsea. A war of words broke out as each side accused the other of unpatriotic actions. The unions argued that the strike was against "un-American conditions," while employers and federal officials claimed the walkout jeopardized the war effort. Aberthaw, one of the largest open shop contractors in the area, plastered their Squantum job site with hortatory posters in English and Italian:

THIS WORK IS MORE THAN A CONSTRUCTION JOB. It is our chance to help win the war. Our work must be done in the same spirit as the work of our friends, brothers, sons who fight the battles. This plant is being built to guard their lives. A delay on your part may mean death to them.[3]

Employer appeals to patriotism were laced with venomous denunciations of those who questioned the military juggernaut. Union critics of wartime labor policies paid a high price. On the national level, Hutcheson and the UBCJA leadership were regularly labeled traitors and slackers. Though other AFL officials privately supported the carpenters, they chose to stay in the background, intimidated by the public abuse that followed Hutcheson's condemnation of the Baker–Gompers agreement. In a later New York ship carpenters strike, United States Shipping Board head Edwin Hurley charged Hutcheson with "adding to the fearful danger our soldiers already face." Influential preacher Billy Sunday whipped antiunion hysteria to a higher plane, invoking the name of God to denounce the UBCJA chief's treason.[4]

Boston-area unionists gritted their teeth and resisted similar indictments in the local media. John McDonald, secretary of the Building Trades Council, angrily complained that contractors were exploiting the workmen *and* the government by gutting traditionally accepted working conditions and charging exorbitant rates under the phony guise of patriotism: "We are just as patriotic as they are and we believe more so; although we don't wave the American flag continuously or endeavor to use it as a protecting shield to gather in excess profits and break down conditions at the expense of the wage-earners."[5]

A livid Stanley King, assistant to Secretary Baker, arrived in Boston and ordered the men back to work. King told the press that the Boston building trades unions were the only ones in the country to violate the Baker–Gompers compact. King's order was seconded by James Donlin, president of the AFL's Building Trades. Amazingly, the union leadership held firm despite the pressure. Nor did individual workers waver. All the projects remained shut down. The Building Trades Councils voted unanimously to ignore Donlin's telegram and announced pointedly that "any settlement to be made will be made by the joint building trades of Boston and Quincy and no others."[6]

On November 16, the strike ended. The Departments of War and Navy promised to confer on the open shop conditions. The Building Trades Councils declared the men would return to work out of "pure patriotism" on the basis of nothing more than such an assurance. The end of the Boston dispute did not, however, resolve the larger issue. For the next several months, building tradesmen across the country refused to allow the government to sabotage firmly established working conditions. Scattered walkouts impressed the importance of the closed shop on the Wilson administration. The persistence of the strikers and the Brotherhood in particular paid off. In April 1918, the government disbanded the ineffectual tripartite boards and appointed a National War Labor Board. Under the new arrangements, local conditions were guaranteed. If an area had a closed shop before the war, so it stayed.

The War Labor Board won the approval of organized labor. Labor disputes slowed as unionists discovered that the Board's rulings were surprisingly considerate of their grievances. Even the harshest critics of governmental intervention in labor relations, such as William Hutcheson, accepted the Board's machinery. When several Brotherhood locals requested support for wage strikes in October 1918, the union's general executive board rerouted the appeal to the appropriate federal representatives. Not everyone was equally pleased. Employers were generally less sanguine, fearing that federal sanction of "a living wage" gave unionism too great a boost.

Furthermore, the War Labor Board established an unwanted, if mild, precedent of state involvement in internal business affairs. In a period of opportunities for inflated profits and instant fortunes, employers wanted to pursue their interests unfettered by even half-hearted federal watch-dogging. Shortly after the war, Newton Baker confirmed many unionists' suspicions when he concluded that labor had been "more willing to keep in step than capital."[7]

James Donlin, like Baker a sharp critic of wartime labor militance, spelled out the implications of that double standard.

Wages did not keep pace with the increased cost of commodities. Big business, little business, all business showed a total disregard for the Government in its life and death struggle. Evidently no one expected business to be even reasonable when there was an opportunity to profiteer. Business, more subtle than labor, reaps the riches and avoids the criticism. Still the occasional murmur of the worker was magnified. He was accused of being disloyal.[8]

Arthur Huddell of the Boston Building Trades Council echoed Donlin's evaluation. In the Boston area, he claimed, "profiteering on labor was an outrage and a disgrace."[9]

Though wages rose during the war, prices practically doubled from 1915 to 1919. Nor did the steadily upward climb end with the Armistice. Between June 1919 and June 1920, the cost of living continued to soar at a 22 percent clip. A *Boston Globe* survey showed that from 1914 to 1920 carpenters' wages lagged over 14 percent behind a cost-of-living index.[10] Federal policy had stabilized production during the war, but the Wilson administration showed little interest in smoothing the transition to peacetime. The sudden and complete eradication of military industries severely affected carpenters' fortunes. Private residential construction rebounded in 1919 but not nearly enough to provide jobs for those displaced from the vanishing shipbuilding and military housing sectors.

The echoes of the wartime experience hovered over labor relations in the stormy year of 1919. The trauma of overseas death and destruction and the mounting anger at domestic profiteering provided a backdrop for workers eager to make up for their sacrifices. When the Holyoke Building Trades Employers Association locked out the city's construction workers in May, the unions justified their wage demands by pointing to wildly inflated prices and unremitting rent-gouging. The Building Trades Council saved its most caustic barbs for the employers' attempts to continue building with hungry war veterans at "starvation wages." Sparing the returning heroes the usual invectives hurled at scabs, the Council, sternly lectured the BTEA on the ethics of their strategy: "It must be remembered that these boys were making sacrifices while the employers were making huge profits."[11]

The Holyoke craftsmen were but one splash in a national pool of striking workers. In all, over four million workers went on strike in 1919, an incredible 23 percent of the total labor force. Tens of thousands of workers struck in Massachusetts alone. Textile workers in Lawrence walked out for the eight-hour day with no reduction in pay. Telephone operators in Boston and across New England won a series of demands in April. Striking against the federal government (which still held wartime control over the telephone companies), the operators broadened the labor movement's horizons beyond private-sector unionism. "I do not believe," wrote one observer, "that an industrial issue has ever before penetrated every village, hamlet or town of New England as has this strike of telephone girls."[12] In the state's most famous strike of that year, Boston's policemen organized the nation's first police union. Following the dismissal of nineteen union leaders, the city's police force walked off their jobs in a bitter and protracted battle that ended with the replacement of an entire generation of law enforcement officers.

The strike wave was not only a matter of immense numbers, but of unprecedented scope. Every section of the workforce took part and, as historian David Montgomery has written, "every conceivable type of demand was raised"—improved wages, reduced hours, union and shop committee recognition, joint negotiating councils, defiance of governmental decrees, and freedom for jailed unionists.[13] The remarkable general strike in Seattle involved 110 local unions (many with binding contracts) supporting the city's locked-out shipyard workers. The Seattle General Strike Committee, representing thirty-five thousand workers, took responsibility for the delivery of food supplies, health care, and public order, virtually functioning as a substitute city government. Realizing the action's historic significance, the union-owned *Seattle Union Record* raised the possible implications of such a general strike.

Labor will not only SHUT DOWN the industries, but Labor will REOPEN, under the management of the appropriate trades, such activities as are needed to preserve public health and public peace. If the strike continues, Labor may feel led to avoid public suffering by reopening more and more activities, UNDER ITS OWN MANAGEMENT. And that is why we say that we are starting on a road that leads—NO ONE KNOWS WHERE![14]

It was a time of heightened political consciousness and solidarity within the labor movement as a whole. Skilled craft workers shed some of their aloofness toward semi-skilled and unskilled workers. During the New Bedford textile strike, the local Building Trades Council voted to stop all construction work on the mills if the mill owners brought in scabs. Brockton carpenters won some extraordinary contract language, proving that questions of workers' control were not limited to the Seattle strike. After three weeks on strike, the Brockton men won a 15-cent-an-hour increase, a shorter work week, and an astonishing clause that read: "The union also reserves the right to restrict the amount of profit the master carpenters shall receive from the work of each carpenter." A joint union-master committee was put in place to establish the maximum figure and monitor the agreement. This step limited potential profiteering and injected the working carpenter directly into the heart of the building industry.[15]

In April, May, and June, carpenters struck in Arlington, Burlington, Melrose, Natick, Needham, Newton, North Attleboro, Reading, Stoneham, Wakefield, Watertown, Wellesley, Wilmington, Winchester, and Woburn. But the biggest strike was left to the carpenters of Boston. In early May, carpenters in the thirty locals affiliated with the Boston Carpenters District Council voted 6,000 to 700 to strike, if necessary, for a jump in wages from 75 cents to $1 an hour and a forty-hour week. The Boston Building Trades Employers Association rejected the union position, but as the strike deadline neared, most of the non-BTEA employers bolted. On May 14, 4,500 carpenters won the demand outright, requiring only 1,400 others to walk out.[16]

The 250-member BTEA (formed in 1916 out of those MBA employers who were prepared to bargain with the unions) counted on the general postwar construction downturn to sober the striking carpenters. Discouraged by the large number of builders who granted the union demands, the BTEA nevertheless understood that an unyielding stance on their part would eventually force non-BTEA contractors to retract the wage advance. In a series of statements, the Association outlined its position. A 33 percent pay increase would only act, they argued, as a "hindrance to the resumption of work in the building trades." They claimed the wartime raises were more than sufficient since the buying public still needed "time to adjust itself" to the higher costs of building. The Association further accused the unions of nipping a

"renewal of confidence" in the bud by deterring those owners who had been on the verge of a "partial adjustment." And finally, the BTEA held out little hope for union success. "It is almost a foregone conclusion that the big issue in breaking the strike will be the unprecedented scarcity of construction work, coupled with the excessive increase in wages demanded under such adverse circumstances."[17]

Market conditions favored the contractors. In past disputes, carpenters had generally succeeded in booms and failed in slumps. But 1919 was not an ordinary year and reliance on ordinary tactics proved fruitless. The BTEA had not considered the reinvigorated solidarity sweeping the labor movement. They expected the squabbling between building trades to continue right through the carpenters' strike. They had good reason. Four years earlier, the carpenters, lathers, and plasterers had pulled out of the AFL-affiliated Boston Building Trades Council and reemerged as the rival Allied Building Trades Council. Thus, when the carpenters struck in May 1919, the employers expected sympathetic support would only come, if at all, from the two other trades in the ABTC.

The two councils met early in the strike, however, and decided to bury the hatchet. They agreed that if a scab appeared on any struck job, union craftsmen in every trade would immediately walk out, precipitating a general construction strike. Support for the carpenters steadily mounted as the dispute continued. The District Council was financially solvent. By the end of May, $25,000 in strike benefits had been distributed. On May 26, the sixteen hundred members of the UBC shop and mill locals in the Boston area joined the strike. That same day, the thirty-five building trades of the city united under a new charter from the AFL in the United Building Trades Council. The fledgling organization declared its intention to "work as a unit in combatting any attempts of the employers to change the conditions of work" and voted additional funds to support the strikers. On June 2, the plasterers laid down their trowels, bringing the total number of building trades workers on strike to four thousand.[18]

The formation of the United Building Trades Council threatened an even greater impasse. In early June, the Council voted to call out its thirty thousand members if an agreement was not forthcoming. The prospects of a general strike in Boston brought Mayor Andrew Peters into the fracas. After two weeks of negotiations, Peters convinced the BTEA to accept a novel compromise authored by UBCJA International Vice-President T. M. Guerin. The agreement covered all thirty thousand tradesmen in the UBTC and provided for a standard wage for *all* skilled construction tradesmen—90 cents an hour immediately and $1 an hour beginning April 1, 1920.

From the union perspective, the standard wage encouraged cooperation and unity among the trades; from the employer vantage point, the common scale removed one of the motivations (the desire to win higher-paid work) for jurisdictional disputes. Both groups hoped the uniform conditions and the ensuing stability would help revive the slipping market. For the duration of the settlement—until December 31, 1920—Boston's construction workers were granted a closed shop once again. In turn, they agreed to submit all disputes to a board of arbitration. It was a major union victory, cementing union control in the industry and advancing solidarity between the trades. The *Carpenter* hailed the contract as "one of the most significant documents of the kind ever signed."[19]

9

Cooperatives:
Building Without Bosses

During the spring of 1921, a group of Boston area tradesmen traveled daily to the South Shore town of Quincy. They were building a new $8,000 house for a Boston journalist. The house was nothing special—a small bungalow, half wood and half brick on the outside. But inside, as the job neared completion, the tilesetters discreetly placed a number of bluish-green tiles above the mantelpiece. The tiles contained the names and local unions of the men who built the house along with a small design of characteristic tools of each trade. The house was, according to John Nason, "the first product of a thousand cooperative working builders, with the boss eliminated and with democratic management substituted."[1]

Nason was the president and business manager of the Building Trades Unions Construction and Housing Council, a Boston building cooperative incorporated in late 1920 at the urging of several members of the Boston Bricklayers' Union. Shortly after the war, that union had appointed a committee to establish contract bargaining guidelines. After months of discussions, the group concluded that continually escalating building prices functioned as an obstacle to further wage increases. Em-

ployers forcefully argued that labor costs had to be held in check before overall costs could stabilize. Unionists disagreed, but realized that they had to counter this argument effectively in order to win raises. The Bricklayers' committee claimed that the price of labor had little impact on spiraling costs. Instead, they blamed contractor and subcontractor overhead.

Nason and another committee member named George Edwards believed that the unions had to do more than present their case rhetorically. If, as they suggested, excessive employer overhead produced overpriced buildings, builders who disdained profiteering should be able to reduce construction costs substantially. But few conventional builders volunteered to cut their profit margins. "We determined," Nason told a Boston Chamber of Commerce hearing, "that we ourselves were more competent to carry the thing on and reduce the cost than could be done under the general contract system."[2] After laying bricks for eight hours a day, Nason and Edwards spent their evenings lobbying building trades union officials to support the establishment of a union-run building cooperative.

Their proposal was not a completely new

idea. Nineteenth-century America had seen waves of producer cooperatives come and go. The Knights of Labor considered cooperatives the seedbeds of a new society. Terence Powderly, head of the Knights, had long favored the creation of cooperatives over other trade union tactics, such as strikes and boycotts. Powderly believed that the K of L cooperatives would expand and flourish, taking on more and more essential productive tasks until they eventually replaced the existing wage system. In 1887, the Knights sponsored 135 producer co-ops including boot and shoe companies in Lynn and Spencer, Massachusetts. The AFL had never put much stock in these worker-ownership schemes, preferring to improve the lot of union members through collective bargaining with employers. But cooperatives were not an entirely foreign concept for AFL-affiliates either. The UBCJA, for example, set up a short-lived Northwestern Cooperative Building Association in Minneapolis in 1886.[3]

Brotherhood carpenters in Massachusetts had their own extensive experiences with worker-initiated firms. Striking carpenters had frequently used the competitive threat of a union-operated building company to bring reluctant contractors back to the bargaining table. In some cases, the unions simply allowed individual members to approach the public as contractors, as they did in the strikes in Fall River in 1900 and Pittsfield in 1906. In others, the unions organized full-fledged cooperatives, ready to capitalize on the skills of the membership for large-scale projects, as they did in Lawrence in 1900 and Springfield in 1904. The practice continued into the 1920s. When builders in Worcester locked out the city's construction workers in April 1921 in order to impose a 20 percent reduction, the Carpenters District Council immediately authorized twenty members to set up their own businesses. Less than a week later, a joint meeting of all the building trades announced plans to form a Builders' Cooperative Society

that would "furnish employment to hundreds of members."[4] In each case, the union's public stance was identical. *We* are the builders, they would announce. We know the work and how to do it. We will demonstrate that the contractors can afford our wage requests by performing carpentry at the proposed higher wage rate without increasing the consumer's bill.

These tactics proved useful for a number of reasons. First, if worker-initiated building got off the ground, it provided extra income for striking carpenters and saved the locals the expense of strike benefits. Second, it was an effective method to win public approval for union demands. Third, the threat of union competition often did force the employers to reconsider their negotiating position. And finally, it was a morale booster for the men. It reminded them of the crucial role they played in the industry. The unions' public statements were not overblown propaganda. The workers *were* the builders. The fact that they could provide quality construction without the contractors reinforced self-respect and pride in work.

The Nason and Edwards scheme initially piqued interest for these same tactical reasons. In the midst of the 1919 strike, the two enthusiastic bricklayers convinced the Boston Building Trades Council to support a plan for a building cooperative to construct "up to date and artistic houses for workers." The BTC leaders' search for every strategic advantage during the strike momentarily dovetailed with Nason's and Edward's larger social vision. On May 29, the Council announced that it intended to purchase a 12 acre tract between West Roxbury and Dedham in order to build eighty houses.[5] The plans were not carried out, and council leaders may have never intended to use the idea for anything more than a bargaining chip. The strike ended successfully for the unions. The men returned to work, and the grandiose co-op structure temporarily slipped between the cracks.

Nason and Edwards refused to let the matter die. Meeting weekly with other interested unionists in a small room at Wells Memorial Labor Temple, they enlisted the aid of Harvard economist Edmund Lincoln to develop a feasible and financially sophisticated blueprint for the cooperative. They decided to go to the general membership of the various trades—some twenty thousand workers—and ask for a $1 contribution per person. With that initial capital, they hoped to launch a cooperative bank that would grant construction loans to workers in need of housing. The worker would then pay off the loan in rent with an ultimate option to buy. According to the prospectus, the financing served to provide start-up capital, boost housing demand, and alter economic priorities such that "the worker should have the right to work and use the land and also to receive the full return of his own toil."[6]

The authors had high hopes, lofty goals, and boundless enthusiasm for the project. They buttonholed union leaders, journalists, and local politicians to promote the scheme. One sympathetic account in the *Boston Globe* described the Council as "one of the most unusual undertakings of its kind ever launched in the world" and predicted that it might be an "incentive for a widespread progressive labor movement involving labor's general participation in the management of industrial enterprises of all classes and kinds." The Boston cooperators received encouragement from similar enterprises across the country. The St. Louis and Minneapolis Building Trades Councils established their own contracting companies. Building trades delegates to the 1921 Illinois AFL convention won the unanimous passage of a resolution calling for a statewide building cooperative. "The workers in the building trades are quite capable of conducting the building industry," read the wording of the resolution. "With the co-operation and loyalty of every trade unionist and co-operator in the State it would only be the matter of a few years

when the Building Trades of this State would be the masters of their own destiny, and the builders of most of the buildings in the State."[7]

Undertakings packaged in such grandiose terms rarely meet the impossibly high expectations they foster. The Boston Construction and Housing Council was no exception. Still, by the end of 1920 almost $10,000 had been raised and a rough organization was in place. Nason took the Council's one full-time salaried position at the bricklayers standard union rate. During its first year, the Council completed contracts for two houses, a number of garages, and extensive minor home repairs. Despite the stated goal of providing steady year-round work, the cooperative was forced to hire by the job, giving priority to Council members. The lowest number of men on the payroll was sixteen, but by mid-winter 1921–22 the total number of employees had increased to twenty-eight.[8]

The cooperative was meant to be more than another name on the bottom of a worker's paycheck. Though a few of the charter members withdrew and returned to conventional construction, the majority relished the difference. One worker told a researcher:

When I go to a boss and ask for a job, do you suppose he asks me how well I'm trained? Not in a lifetime. What he wants to know is how fast I can work, not how well. I don't want to work that way. I know my trade from A to ampersand; I like it; it's the best work there is if you can have a free hand, and the cooperative gives me that. I want the chance to put my heart into my work, not just to spend so many hours a day doing what somebody tells me to do.[9]

The cooperative was structured to foster those feelings as much as possible. All the estimating, purchasing, and building was done by Council members. The client just provided the architect's plans. On the site, the workers elected a building committee with a representative from each trade and one general foreman. Each craft was responsible for its set of tasks. Supervision was perfunctory. As Nason remarked, "we don't expect that [the foreman] will spend much time supervising. The men are working for themselves, so they don't need that."[10]

For all the positive feelings, the Council was unable to generate sufficient capital funding. Nason reached out to unionists outside the building trades in order to increase the number of shareholders, but still fell short of the money needed to bid on the sizable projects that would have created more employment and more income. In 1922, the Council made a fatal investment error, buying a large piece of land in Dorchester outright. Nason hoped to build a two-story apartment house and sell it quickly in order to shore up the Council's shaky finances. When the inevitable delays occurred, the cooperative ran short on cash and missed several payrolls. The business never recovered, limping along until the Depression, when almost every marginal building firm was destroyed.

The Building Trades Unions Construction and Housing Council was a noble, if quixotic, experiment. Cooperative members got a taste of a worker-initiated, worker-controlled, democratic operation. Its very existence was premised on the irrefutable fact that building tradesmen had all the tools, literally and figuratively, to conduct work on their own. Sixty years later, after decades of further deskilling and specialization in the industry, the notion that construction craftsmen have the wherewithal to "build without bosses" still has a powerful appeal. In 1984, tradesmen in Buffalo, New York, formed the Buffalo Building Trades Council Construction Corporation, a nonprofit union-directed building firm, as bargaining leverage during difficult contract talks. Unlike Nason's and Edwards's Council, the Buffalo company was strictly a negotiating ploy, to be dropped in the event of a satisfactory agreement. Like the experiment of the twenties, however, the Buffalo formation also rested on the assumption that building trades workers can construct the buildings of America by themselves. That collective sense of self was and is not common among American workers, yet it has been a recurring theme for skilled construction workers. The Boston cooperative removed the concept of workers' control from the abstract into the realm of reality. Though it failed, the brief history of the Council stands as a concrete monument to Nason's dream that a productive enterprise could "be run by men and not by dollars."[11]

Everett Street bridge in Allston, Boston and Albany Railroad, 1897. SPNEA, B&A

Houses, Factories, Ships,
and Railroads,
1897–1925

Construction crew on the B&A Railroad in Newton. SPNEA, B&A

Railroad bridge in Brookline, along the Highland branch of the B&A Railroad, c. 1900. SPNEA, B&A

Under the boss's watchful eye on a bridge in Natick, B&A Railroad, 1907. SPNEA, B&A

Pittsfield railroad station site, 1913. SPNEA, B&A

Outside the Pittsfield station, 1913. SPNEA, B&A

Scaling multistory ladders in the Pittsfield station, 1914. SPNEA, B&A

Textile factory under construction, part of the giant Pacific Mills in Lawrence, 1910. MATH

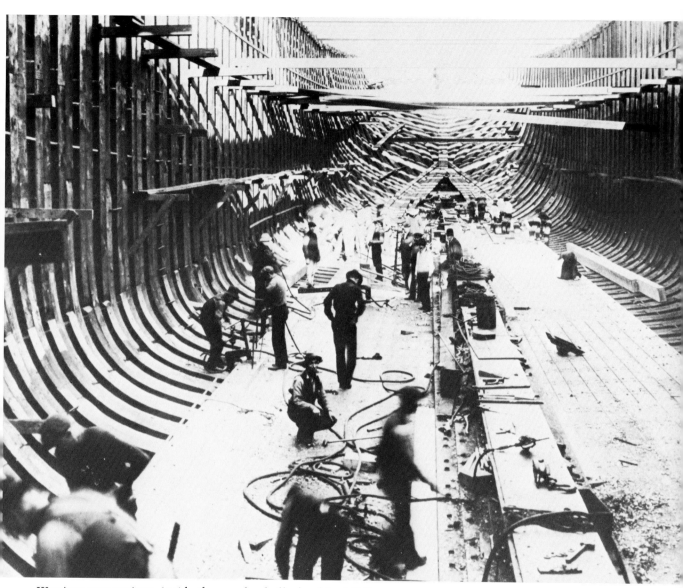

Wartime preparations, inside the wooden hull of a ship, 1917. NA

Building site on Empire Street in Allston, 1925. (*Above*) Man in front, with pipe, is Hubert Greenland, father of John Greenland (past president of Carpenters Local 40). (*Below, from left to right*) Isaac Greenland, Patrick McKenna, Cecil Greenland. John Greenland

10

The American Plan

The 1919 strike wave frightened employers. Coming on the heels of wartime labor shortages and federal support for union pay scales, the specter of a relentlessly organizing labor force haunted the captains of American industry. Union membership reached five million in 1920, twice the number of unionized workers just four years earlier. As a steel mill owner commented, it seemed as if "the gods were fighting on the side of Labor." But the heavens and labor soon parted company. Bruised but not battered, business leaders across the country retooled for a new offensive. Federal authorities, no longer constrained by the need for wartime labor peace, launched an antiradical political vendetta to complement the antilabor campaigns. Employer associations of all kinds rediscovered the value of class solidarity, organizing a furious and unceasing assault on union labor under the label of the "American Plan."

Building employers, suitably underwritten by industrialists and local Chambers of Commerce, did their share. They tackled construction unions in San Francisco, Chicago, Milwaukee, Detroit, Minneapolis, and Philadelphia. In San Francisco, the Industrial Association broke the twenty-year reign of one of the

country's mightiest closed shops. Financed to the tune of $1.25 million, the Industrial Association overpowered the unions and prohibited collective bargaining in the city's building industry after June 13, 1921. An Association pamphlet labeled the practice a violation of "the principles of the community."[1]

Builders had particular axes to grind in the postwar turnabout on labor. Beyond their obvious objections to the numerical growth of the building trades unions, contractors complained of a cumulative loss of control and authority on the job site. In the supportive political climate of the American Plan era, building employers felt free to propose sweeping changes. For example, a column in the *The Builders' Record,* the Boston BTEA monthly, ticked off a list of "essential requirements for a successful year in building construction." These modifications added up to a repudiation of the fundamental tenets of unionism in the industry. BTEA suggestions included the elimination of any restrictions on output, a ban on sympathetic strikes and jurisdictional disputes, absolute freedom in hiring and firing, removal of the right of a union representative to consult with workers during working hours, and the transfer of the foreman from the obligatory status of union member to unquestioned agent of the employer. BTEA members attributed the increased costs of buildings to the many layers of union rules. These rules, fumed Arthur Joslin of Joslin & Landry, "do not leave it to our or our foremen's judgment as to just how or who will do a certain class of work."[2]

The image of a shackled and helpless foreman symbolized much of the contractors' indignation. Most locals' by-laws required foremen to be union members. These crucial clauses did, in fact, soften ties to the employer and presumably preserved the foreman's loyalty to the union. Few foremen objected to the rule. Given erratic construction manpower needs, building employers offered foremen barely more job security than any other employee. They rarely held permanent supervisory positions, instead floating back and forth from foreman to working craftsman. As long as a foreman knew he might return to the tools one day, he was reluctant to turn his back on the union and cast his lot with an employer. "It is getting increasingly difficult," moaned army construction chief Charles Gow, "to find a foreman who regards his obligations to his employer as superior to his union."[3] Even more importantly from the contractors' perspective, foremen hesitated to be severe disciplinarians, knowing that yesterday's underling could easily be today's coworker and, possibly, tomorrow's boss. As union members, foremen were represented by their locals in collective bargaining. Their rates of pay were often negotiated individually with employers, but always in relation to the standard union scale. And if all these bonds were not enough, union by-laws included stiff fines and/or the penalty of expulsion for foremen who violated work rules. Regulations for carpenter foremen in Boston stated: "they shall not rush, use abusive language or otherwise abuse workmen under their direction."[4]

This notion of a "foreman on a leash" was firmly embedded in contractors' minds. On occasion, foremen played on this concern in order to evade responsibility for poor production. In a report filed with a major Boston construction firm, one nervous foreman justified slow job progress by his inability to supply the necessary aggressive leadership for fear of "getting in wrong with [the] Local and having a fine put on." In closed shop towns, a fine was the least of a pusher foreman's worries. On the giant $24 million Boston Army Base job during World War I, a notoriously offensive foreman had his union card taken away by his local on the charge of profanity. Though he was the contractor's favorite foreman, widely feared by the men for his hard-driving supervision, he was still discharged. Much to his chargin, the contractor could not retain the offender since the man was no longer a union member and it was a closed shop project. All in all, Joslin spoke for most employers when he lamented, "foremen very seldom get after the men the way they ought to."[5]

BTEA employers also linked the shorter work week to their loss of authority. By 1921, five of the twenty-seven building trades (including the carpenters) worked forty-hour weeks. The rest were obliged to show up on Saturday mornings for four additional hours. The lack of uniformity created problems. As contractor Ralph Stewart pointed out, it was harder to organize the extra four hours of work without the presence of carpenters. Employers therefore often chose between delaying carpentry tasks or paying double-time for overtime hours. According to Gow, either option was unacceptable. In the absence of crucial carpentry work, Gow claimed Saturday mornings were unproductive and turned a forty-four-hour week into a *de facto* forty-hour week in terms of output. Paying overtime for just one craft, on the other hand, caused dissension between trades and ensured the eventual implementation of a shorter week for everyone.[6]

Contractors blamed their "predicament" on

the war. As overall construction slowed in 1920–21, the massive federal contracts and accompanying record-setting profits of the war years receded from their memories. As the builders looked back, the clarity of their vision was clouded by their obsession with labor's gains. Gow could only recall the period as a time of excessive worker power. Steady year-round work had been, for once, available, and tradesmen sometimes selected between competing jobs. Stewart believed that the combination of ample work and standard union wage rates had driven worker efficiency to an all-time low. With large numbers of new and untrained "green hands" receiving the same pay as experienced union craftsmen, Stewart suggested that "the efficiency of labor [was] measured by the most inefficient" worker. Joslin asserted that the efficiency level of a postwar carpenter was half that of a prewar man. Employers recited as gospel the conclusions of a 1921 construction report from the War Department to support their contentions:

The universally attractive high standard wages paid to organized labor have placed the second-rate craftsmen on a par with the high-class efficient artisans; and instead of the average day's work being raised, it is proportionately lowered because the first-class journeyman must carry along his less efficient brother. [7]

Labor officials acknowledged the unusual nature of the wartime job site environment. Year-round work was a novel experience for most construction workers and the scale of jobs was unlike anything seen before. The Boston Army Base project had 7,500 workers, two-thirds of them skilled mechanics. The government paid little attention to total expense, preferring to contract on a cost-plus basis rather than to hold builders to fixed estimates. Federal priorities emphasized the total volume of work regardless of the size of the workforce or the number of overtime hours worked. Even

John Carroll of the Boston Cement Finishers Union argued that the wartime system offered "no incentive on the part of the supervisory force to get the desired efficiency out of the men." [8]

The contractors had never noticed or, in any case, mentioned these problems during the war. On the contrary, the cost-plus system combined with minimal federal oversight meant that whatever costs a contractor declared were quickly reimbursed with little or no inspection. A certain profit margin was guaranteed regardless of the efficiency of the operation, and that margin could be boosted easily with minor manipulations of the figures on labor and materials. It was only later, when the federal cookie jar closed, that builders developed a conscience about bad wartime labor habits.

Building tradesmen had no quarrel with the employers' version of the nature of the wartime job site, but they deeply resented any attempts to pin the blame on lazy and inefficient workers. Arthur Huddell of the Boston Building Trades Council angrily reminded contractors that they had been the ones to pressure the unions to accept a horde of contractors' relatives and acquaintances. These men were "not mechanics and did not pretend to be mechanics," said Huddell. Furthermore, he noted, the contractors did not care, since "Uncle Sam paid the bills." As a result, the employers set an overall lackadaisical tone for production.

The feeling among the men was that there was plenty of time for the work to be done. Two men were set to do one man's work. . . . They worked side by side, the mechanic and the man who knew nothing about the business. . . . Now you come along and want efficiency. Who broke down the efficiency of labor? I say the employer and the cost-plus system. [9]

Huddell and other Boston building trades

workers realized that the heightened concern over inefficiency and "excessive" union power was the first step in an employer campaign to capture the hearts and minds of the public in an upcoming showdown between management and labor in construction. The American Plan–vintage employer was a much more sophisticated operator than his predecessor. Boston contractors knew they had paid a political price for their previous standoffish and arrogant stance in labor disputes. Charles Gow told a Boston Chamber of Commerce hearing of his frustration with a tradition of public support for construction union craftsmen against their employers. It was a tradition that the BTEA was intent on reversing.[10]

The city's builders had been licking their wounds since the 1919 agreement. By the end of 1920, conditions appeared increasingly favorable for a counterattack. Building had declined even further; workers were desperate for employment. The official unemployment rate for construction workers in Massachusetts in the winter of 1920–21 was 27.6 percent, the highest since the depression of the 1890s. Despite their own lack of contracts, a number of the state's builders actually welcomed the salutary influence of postwar unemployment, suggesting the revival of economic insecurity had "resulted in an entire reversal of attitude on the part of the average employee."[11] In addition, spiraling inflation was finally slowing, enabling employers to argue against union demands for wage increases to keep up with the cost of living. From the BTEA's point of view, the economic circumstances and political climate of 1920–21 could not have been more ideal for a siege of the unions.

The 1919 agreement expired on New Year's Day 1921. Union members sought a raise. Though prices had stabilized since the summer of 1920, building trades workers felt overall that the two-year agreement had put them even further behind in their standard of living. "Somebody says the cost of living has come down 10 percent," said Arthur Huddell. "But it went up 160 percent. Labor never got a 160 percent increase in wages." The wartime focus on military construction had drained resources from the residential market. The ensuing housing shortages were driving up workers' rents astronomically. Thus, despite the larger postwar paychecks, most rank-and-filers agreed with Huddell and James Dobbs of Plasterers Local 10. "I would prefer to go to the 1914 rate of wages, 65 cents per hour," Dobbs said, "and have the dollar buy now what it would buy in 1914." In the early winter, the United Building Trades Council informed the BTEA that the city's tradesmen wanted an increase from $1 an hour to $1.50.[12]

The BTEA refused immediately. On December 29, the Association offered instead a wage freeze and a cut in overtime pay from double time to time-and-a-half. Faced with the cold winds of a New England winter on top of a slack building year, the UBTC realized the precariousness of its bargaining position. On January 10, the unions proposed extending the 1919 agreement until May 1. As the Council backpedaled, the employers quickly moved in for the kill. On January 13, the BTEA announced a new set of working conditions to govern job sites beginning on January 20. Under the new regime, wages would be cut by 10 percent to 90 cents an hour, overtime would be paid at time-and-a-half, a uniform forty-hour week would be established, and the travel time formula was to be amended in such a way as to add an unpaid hour and a half to the working day.[13]

Union officials were stunned by the unilateral action. They had assumed the extension of the old contract would be automatically accepted. UBTC spokesmen John Carroll and E. A. Johnson reported that the BTEA announcement came as a complete surprise. Scrambling for an appropriate response, the Council issued a statement denouncing the ultimatum as a threat to "the principles of collective bargaining, decent travelling conditions, and humane laws in trade organizations." The unions knew the employers were capitalizing on the cyclical and seasonal slowdown, allowing them to take a strike "without any financial loss." The BTEA offer still incorporated the principle of a closed shop, but slump or not, the concessions were unacceptable to the unions. On the twentieth, ten thousand construction workers within a 45-mile radius of the city of Boston walked off all BTEA-member projects.[14]

The strike delighted the BTEA. According to the *Boston Globe*, the Association's resources were well up into the hundreds of millions of dollars. "There never was a time in the industry," crowed BTEA secretary Joseph Walsh, "when the employers were so close together." The MBA, alienated from the BTEA since the 1916 split, rushed to provide assistance and close ranks. The employers took a leaf from the unions' book, organizing and expanding instead of just waiting the workers out. Within a week of the strike, the Association added thirty-six new members and organized two chapters outside the city. They assured the public they were not interested in crushing unionism but merely intent on reforming a labor agreement that "blocks progress and penalizes initiative and kills ambition." They discussed plans with and won endorsements from independent contractors, real estate dealers, and architects. And they even adopted the traditional union form of the mass meeting, firing up five-hundred Association members at the plush City Club to a level of frenzy that "shook the chandeliers." When inspirational methods failed to do the job, the BTEA turned to hardball. A number of public officials accused the Association of coercion and collusion. James Moriarty, president of the Boston City Council as well as a business agent for the Sheet Metal Workers Union, claimed BTEA leaders forced uncooperative builders to join or be "locked out" from any future contract work in the city. The president of the United Build-

ing Trades Council wryly remarked that if he had employed similar methods it "would have resulted in 10 indictments."[15]

The strike continued with no signs of scabs or returning union craftsmen. A few smaller non-BTEA jobs employed union workers, but the dominoes did not fall as they had in previous years. On the contrary, not only did Association employers not give in on a one-by-one basis, instead the association pulled in independent builders and thereby removed a significant number of jobs for union men. By February, the original UBTC demand for a wage increase was ancient history. The unions were fighting to stave off reductions. "Labor go back? To what? The Poor House?" asked Arthur Huddell in frustration. "If a man is going to starve to death, it is better for him to starve to death loafing than to wear himself out like a dog for nothing."[16]

The employers turned up the pressure. The MBA, the Real Estate Exchange, and the Society of Architects introduced two pieces of related legislation at the State House. One bill would have prevented any interference with employment practices or restriction on output and the other called for the repeal of an existing law requiring any advertisement for scabs to include notification of a labor dispute. The BTEA was convinced that the strike was over and all that remained was to fine-tune the new arrangements. The Association confidently rejected offers of mediation from the Chamber of Commerce, Mayor Peters, and the State Board of Conciliation and Arbitration. They refused to consider negotiating until the "hysteria" over wage reductions dissipated. On February 23, the BTEA asserted that, with twenty-four thousand tradesmen out of work, "Every red-blooded man is tired of loafing and sick of the strike business and wants to go to work now."[17]

They misjudged the mood. After two months on strike, a union referendum revealed that workers rejected the $1 wage freeze by a 15,298 to 2,569 vote, let alone the BTEA-proposed 10 percent cut. But worker militance had no outlet. The unions struggled to devise a counterstrategy. The strike had essentially become a lockout. Its continuation benefited the employers far more than the workers. The usual tactics, such as out-of-town jobs, union-initiated cooperatives, regular mass meetings, and displays of inter-trade solidarity were either unavailable, neglected, or sorely lacking. The first major mass meeting, initiated by the Trade Union Defense Committee of Greater Boston, was not held until late April. Worse, the strong bonds between the trades of 1919 were collapsing. The bricklayers, plumbers, plasterers, and electricians had either withdrawn or been suspended from the UBTC. Cognizant of their weakness, the unions had pinned all their hopes on the intervention of the State Board of Conciliation and Arbitration. When the employers ignored the Board's letters, the Council was reduced to pleading—without success—for the $1 wage rate.[18]

In other strikes, the employers had been divided over the issue of compromise with the unions. In 1921, whatever dissension existed flowed in the opposite direction. Once the 10 percent cut had been proposed, not a single employer could be found in favor of the original wage freeze offer. The *Boston Globe* reported that if not for a small group of BTEA members (mostly former unionists themselves) who favored keeping the closed shop, the Association would have crushed the strike and declared an open shop. On March 11, the BTEA withdrew its previous proposal and announced plans to deal directly with the workmen, not the UBTC. A Cambridge job site was opened with police protection. In the next two weeks, twenty-five small to medium-sized projects started up again. By mid-April, the closed shop advocates within the BTEA had been isolated. The Association announced the opening of its own Employment Bureau and made plans to conduct business on an open shop basis. The BTEA severed all formal contacts with the

UBTC and put three thousand tradesmen back to work in the following months.[19]

In June, the city's building trades unions acknowledged the new reality and quietly advised their members to take whatever jobs they could find and win the best terms possible. The BTEA set the prevailing wage at 90 cents and fined any member who paid more. "It is only fair to state," *The Builders' Record* reported at the end of the summer of 1921, "that practically all construction jobs are now fully manned at the rates and on the conditions set forth by the Building Trades Employers' Association." The unions never publicly admitted defeat. The strike continued on paper, but as an editorial in the *Boston Herald* commented, it was "tacitly understood by all parties that the strike was lost."[20]

The BTEA had won, but they stopped short of publicly labeling Boston an open shop town. In open shop Los Angeles, employers paraded their antiunion sentiments under the banner of "Americanism, freedom, prosperity." Boston was different. Mindful of the minority of pro-closed shop attitudes in their own ranks and reluctant to open a Pandora's box with the unions, the BTEA operated open shop jobs without calling attention to the fact. No agreement existed and the contractors were under no written obligation to abide by union regulations. In those rare instances when employers located suitable nonunion workers, they worked unimpeded beside union members. Construction in the city proceeded with little conflict. A December letter to the *Herald* from Ernest Parsons reflected employer assumptions about the changed conditions: "It is very doubtful if we need any more 'agreements' with unions. We have managed to get along very well for several months without any 'agreements' and it is better that way."[21]

Despite the employer triumph, labor relations in Boston's construction industry never approached the completeness of Los Angeles's or San Francisco's open shop. The many years of union organizing paid off even in the dark days of 1921. The BTEA was able to impose its will and unilaterally set working conditions, but it still depended primarily on union craftsmen to put up the buildings. The city's contractors had learned how to maneuver politically, how to win greater public approval, and how to brake the steady stream of union gains. But they were unable to break the union within their employees. Except for a brief initial spurt, the Employment Bureau proved to be a failure and BTEA members hired those whom they had always hired—skilled union craftsmen. Loyal to their trade organizations, Boston's building tradesmen labored without a union contract but with a union consciousness.

When the building slowdown finally eased in 1922, the unions resolved to regroup and rebuild. Construction picked up and the summer actually held out the prospects of skilled labor shortages. With the help of Mayor James Michael Curley, a new contract was signed on July 14. The forty-four-hour week was still in place (except for plasterers and painters) and dreams of a $1.50 wage were long gone. Bricklayers and plasterers boosted their pay to $1.12½ while the rest of the trades returned to their prestrike $1 an hour. Overtime was set at time-and-a-half and, according to the BTEA, a number of "wasteful" practices were eliminated. But more important, the glue that held the unions together—the closed shop and many of the work rules—was still intact.[22]

The eighteen-month strike colored collective bargaining for the rest of the decade. The threat of the open shop hung over every nego-tiating session. A month before the 1922 agreement expired in the spring of 1923, Boston carpenters voted to demand a wage increase to $1.12½, double time, and a forty-hour-week. The BTEA immediately blasted the union's position as a "return to the old, discarded, obsolete, extravagant working rules and onerous, unreasonable trade union conditions of the past." Secretary Walsh ominously warned that "nothing but the 'open shop' with unstabilized conditions looms unless the men's representatives recede."[23]

Once again, Mayor Curley stepped in to ward off a confrontation. In May a settlement was reached. The carpenters won a wage increase—$1.10 for 1923 and $1.12½ for the following year—but lost on the hours and overtime. The closed shop remained sacrosanct. With each succeeding contract, the year and a half of unrestricted employer control slipped more and more into the past. The unions had weathered the open shop storm. They had lost some valuable possessions overboard but managed to maintain an overall forward-moving course. The day after the 1923 settlement, a *Herald* headline wondered "Is It Good News or Bad?" The newspaper's skeptical columnist questioned if the BTEA's high-powered offensive and open shop threats had really accomplished their aims: "Hardly one of the wasteful practices 'the rules' make possible has been eliminated; hardly one of the methods that 'make work' and 'increased' costs has been abandoned.... Is the signing yesterday good news or bad? We confess we are in doubt. Time will tell."[24]

11

Tragic Towns of New England

The Boston BTEA's open shop drive inspired imitators in other parts of the commonwealth. In a few cases, the employers successfully drove down the scale, but, overall, union control over Massachusett's major construction sites remained unquestioned. The construction unions' grip on labor relations in the industry was, in some cases, stronger than their hold on their own membership. General feelings of loyalty to the unions were unshakable, but the once remarkably high level of participation began to dwindle.

The decreasing involvement did not flow from any particular set of choices made by the United Brotherhood or any of the other building trades unions. It was part and parcel of a basic reorientation of American cultural priorities of the era. The labor movement proved to be no more immune to the political complacency of the "Roaring Twenties" than any other institution. Membership in the AFL dropped by one and a half million during the decade. The Federation "shifted from militancy to respectability," wrote historian Irving Bernstein. "With business supreme, the AFL sought to sell itself as an auxiliary of business." This "business unionism" found a perfect

voice in the leader of the AFL. William Green, an honest but uncharismatic officer from the United Mine Workers, stepped into the presidency after Samuel Gompers's death in 1924. Green shied away from overt class conflict and preferred instead that unions "increasingly concern themselves to see that management policies are efficient."[1]

Neither Green's personal timidity nor the conservatism of other AFL leaders, in themselves, account for the depths of labor's "lean years." Many labor officials believed changing economic conditions made combative unionism obsolete. The twenties were, after all, a decade of growth. The consolidation and implementation of years of technological innovations along with refined managerial techniques based on labor intensification boosted industrial productivity by 64 percent. Dividends doubled and mounting profits prompted business spokesmen to pronounce the arrival of a new and permanent prosperity. Though the accumulating wealth was spread unevenly among social classes, American workers still fared well. Between 1922 and 1929, average real weekly earnings increased nearly 15 percent. Secretary of Labor James Davis told the 1928 national Carpenters Union convention this: "There was never a time before when the worker was as well off as he is today. . . . For the first time in human history there is reason for thinking that involuntary poverty is a thing which can be conquered." Such blithe spirits were not restricted to Republican cabinet officials. Former Socialist party member John Spargo argued that the successes of American capitalism disproved the dire predictions of Marxism and other class-based ideologies. "Here in America at least," Spargo wrote in the tone of optimism that cut across political lines in the twenties, "the industrial system and the economic order resulting from it constitute the best and soundest part of civilization."[2]

The mass availability of electricity, central heating, and indoor sanitation ushered in a new era of creature comforts. A consumer culture based on advertising and installment buying gave workers access for the first time to home appliances, radios, phonographs, and automobiles. "More and more of the activities of living," claimed Robert and Helen Lynd in their famous study of Muncie, Indiana, in the 1920s, "are coming to be strained through the bars of the dollar sign." According to the Lynds, this focus on material acquisitions diminished the importance of local trade unions and a culture of labor solidarity. With their new cars and credit accounts, Muncie's working-class families sought to escape rather than strengthen their working-class identities and communities.[3]

"The social function of the union has disappeared in this day of movies and automobile," wrote the Lynds. Observers of the labor scene in Massachusetts generally agreed. The Worcester *Labor News* blamed poor turnout at union meetings on movies, radios, whist, victrolas, autos, and house parties. Unions scheduled fewer dinners and dances, bowing to the reigning preference for weekend activity—the family car trip. The number of marchers at Labor Day parades steadily de-

clined in the early twenties until most cities in the state dropped the celebration altogether in 1926. By 1929, the only public recognition of the once festive and popular holiday in Worcester was a brief radio address by the business agent of the Carpenters District Council. Locals schemed to reverse the trend with fines for absences or prizes for high levels of attendance at meetings. Carpenters Local 1006 in Worcester held a "Get Together Dinner" for members and wives during which organizer Charles Kimball urged the women present to attend regular union meetings in order to reinvigorate the social and cultural aspects of the union.[4]

For all of organized labor's setbacks in the twenties, the construction unions suffered less than other labor organizations. They continued to serve as the backbone of the AFL nationally and locally. Membership in the UBCJA in Massachusetts stabilized at slightly more than twenty thousand. The strong traditions and continuing skilled character of the building trades acted to undercut the listless unionism most observers noticed. The Lynds, for example, speculated that heightened acquisitiveness and individualism were directly related to the erosion of craft skills in industrial settings. Craft pride, they argued, had been replaced by consumerism as a motive force for workers. Construction, however, still required high levels of skill. To the extent that carpenters and other building trades workers maintained a powerful sense of craft, their attachment to their trade and its cultural trappings were less frayed. They had successfully resisted the open shop because their commitment to their collective identities as union craftsmen transcended contract unionism and the presence or absence of a signed piece of paper.

Certainly, the political atmosphere took its toll on the militancy and solidarity of the building trades. The tremendous show of collective unity between the Boston unions in 1919 was not repeated. One trade after another left the UBTC. The city's Carpenters' locals negotiated their 1923 agreement separately. For the rest of the decade, Boston's carpenters stood outside the Building Trades Council and bargained with their employers alone. The number of strikes dropped sharply. From 1924 to 1928, the UBCJA national office sanctioned just six walkouts in all of Massachusetts. The apparent harmony of the period stemmed from the prevailing climate of labor peace, the relatively steady volume of work, and diminished rank-and-file pressure due to lessened participation.

Nonetheless, it would be a mistake to argue that a spirit of individualism had eradicated the culture of cooperation. Strikes may have been curtailed, but they did not disappear. In an eighteen-month span from 1926 to 1928, hundreds of Boston craftsmen walked off construction sites in sympathetic protests against the presence of nonunion workmen of other trades. Outside activities may have cut into involvement in union affairs, but they never replaced a fundamental loyalty. During the negotiations of 1928 (a year often considered a pinnacle of worker indifference) eight thousand Boston carpenters stood in a line stretching from Hanover Street to Washington Street from ten in the morning until six at night to vote on a set of contract demands.[5]

Putting aside the chronic problem of seasonal unemployment, the national construction job situation in the mid-twenties was as good as it had been in years. "I cannot recall a time when affairs in the building industry were running more satisfactorily," said a pleased William Tracey, secretary of the AFL Building Trades Department, in the summer of 1926. "So far as I know, there is not a single building craftsman in the country who cannot find a job at good wages." Tracey's words rang true in Massachusetts' commercial centers. A Worcester newspaper described 1925 as a "crack a jack" year for building. Carpenters worked steadily during the building seasons in Boston, Springfield, and Worcester. Smaller towns ringing metropolitan centers issued record numbers of building permits. Except for the winter months, construction unemployment figures in Newton and Quincy rarely exceed single-digit rates as residential builders hurried to keep up with automobile-aided Bostonians in search of suburban houses.[6]

Not all of the commonwealth's carpenters were so lucky. Joseph Leitao and his brother arrived in the Fall River–New Bedford area in 1919. Both men had made their livings as carpenters in their native Portugal. "When I came here, jobs for carpenters was very slack. You take anything you can get." Joseph worked off and on in the mills and on local farms. He drove trucks and swung a pick and shovel before finally establishing himself as a union carpenter twenty years later. His brother returned to Portugal after one year, his hopes deflated by the lack of opportunities. The Leitaos's broken English hurt their chances to land jobs, but not nearly as much as their site of debarkation. Fall River was a textile town and the textile towns of Massachusetts had passed their prime. Their residents were feeling the pain of an industry's death throes.

In 1924, a *New Republic* correspondent described Fall River as "a city of misery, want, unemployment, hunger and hopelessness . . . of grim and silent mill executives, of lounging men and unoccupied girls aimlessly walking the streets." Fewer than a dozen of the city's 111 textile mills were in full-time operation. The factories were in disrepair, filled with archaic machinery. Since the nineteenth-century heyday of the industry, Massachusetts' mill owners had opted against further investment and modernization. The mills were "living skeletons," in the vivid words of a Fall River clergyman, "stripped to the bone and sinew and sucked of their blood by greedy stockholders."[7] Few of the closed mills ever reopened their doors. The industry left New England for the sunnier climes of the South. The factory jobs were gone forever along with the construction jobs and other employment that depended so heavily on the fortunes of the textile plants. The flight of textile from Fall River was a harbinger of things to come. Most of the mill towns prospered into the mid-twenties before disaster hit. But when the collapse came, it was complete.

From 1920 to 1922, hundreds of carpenters were drawn to Lawrence. The American Woolen Company had contracted to build the Shawsheen Village and the resulting demand for construction workers boosted the membership of the Lawrence Carpenters local from its normal 400 to 500 to 1,500. But the Shawsheen complex turned out to be the last gasp of Lawrence's mill owners. Within a few years, the industry had crumbled along with all related employment. "On account of conditions in the textile industry," a report at the 1925 Massachusetts State Convention of Carpenters stated, "there has been practically no work in Lawrence this past year." Nor was the impending building depression limited to Lawrence. Speakers at the 1925 convention sharply contradicted William Tracey's rose-colored portrait of construction opportunities. Organizing efforts in Newburyport were dismissed as a waste of time, "as there is no work." The situation on Cape Cod was described as "deplorable."[8] Massachusetts had become a divided state—relative prosperity in nonindustrial areas alongside absolute poverty in the textile and shoe towns.

In the summer of 1927, normally a time of peak employment, the unemployment rate for union construction workers in Fall River stood at 29 percent. In neighboring New Bedford, it was an astronomical 46 percent. "New Bedford had nothing," remembers Leo Coulombe. "There were no mills, nothing going on." The following winter, the rate hit 27 percent statewide for unionized carpenters—a figure that was high but still not outside the normal winter range of 20 to 30 percent. However, the distribution by town reflected the common-wealth's uneven sacrifices. The same survey showed that the jobless rate was 43 percent in Lowell, 48 percent in Lawrence, and 60 percent in Holyoke. According to a less "scientific" but possibly more accurate account, nearly two-thirds of Lowell's organized building trades workers had no work at all between 1926 and 1930. With the exception of a post office, scarcely any new buildings were erected in the entire city. Charity had replaced textile as the leading industry.[9]

"Lowell, Lawrence, New Bedford, Maynard, and Fall River . . . are sad, sad places," said Thomas McMahon, president of the United Textile Workers of America. "There is, perhaps, more destitution and misery and degradation in the mill towns of New England today . . . than anywhere else in the United States." If McMahon had been an officer of a shoe workers' union rather than a textile workers' organization, he might have lengthened his list to include the shoe centers of South Boston, Chelsea, Lynn, Brockton, Stoneham, Haverhill, Newburyport, and Georgetown. Estimated unemployment in Massachusetts' boot and shoe industry reached 65 percent by the end of the decade. In 1930, writer Louis Adamic toured Massachusetts on assignment for *Harper's* magazine. Looking back several years later at the article he had written (aptly titled "Tragic Towns of New England"), Adamic concluded that "what was true" in the late 1920s in the industrial towns of Massachusetts "became largely true by 1932 or 1933 of the entire United States."[10]

"The building trades felt the depression before everyone else," says Manny Weiner, whose father was a member of Local 157. Weiner's observation is a comment on the nature of building. The construction industry is always a barometer of economic trends. Private decisions in executive suites to curtail capital investment immediately lower the number of building permits and contracts for industrial construction. Consumer anxieties over purchasing power are promptly reflected in slumping housing starts. Political panic in the face of an eroding tax base inevitably wipes out plans for new schools, roads, bridges, and other public construction. The years leading up to the Great Depression were no exception to this pattern. The devastation of the shoe and mill towns spread like a cancer throughout the state. In the summer of 1928, the Worcester *Labor News* reported "scores of men at the [Worcester Carpenters District Council] Business Agent's office every morning."[11] By the end of the decade, all of Massachusetts building tradesmen functioned under depression conditions.

"I was born in 1916," jokes Tom Harrington, "born either ten years too soon or ten years too late." Harrington's generation of carpenters confronted a frightening situation. Ready to enter the workforce and support young families, these men instead deferred their dreams, drastically lowered their expectations, and focused on the struggle to survive. "There was no work," Ed Henley states simply. Pittsfield Local 444 reported that 70 percent of its members were out of work in September 1930. Another local estimated that only one in twenty was working in November 1931. In the Boston area, between two and three thousand of the twenty thousand building trades workers were on the job in March 1933. Minutes from the Springfield Carpenters District Council indicate that eight to nine hundred of the Council's twelve hundred members were unemployed in May of 1934.[12]

The depression touched everyone in the industry, young and old, experienced and inexperienced. Like a raging forest fire, the depression was out of control, burning everyone in its path. Ellis Blomquist's father was one of many ruined in the bank crash of 1929. "He was jobbing. He was working for himself with a few other guys. He had three houses going and the banks closed him down. It really broke his heart." Today, Leo Coulombe shakes his head and wonders just how the Depression-era carpenters did survive. "You just couldn't buy a construction job," he mutters.

Many carpenters did what carpenters have always done. They tramped. Coulombe went to New Jersey and managed to find short stretches of work on the Pulaski Highway, a Du Pont factory, and a sewage plant. Chester Sewell pursued a similar strategy. "You just had to keep traveling. A hundred miles a day. Stop at a job and ask if they needed help." Tom Rickard's father regularly gathered a few friends and drove off for a week at a time, sharing gas and food expenses in search of work. Ernest Landry's father worked as a maintenance carpenter for General Electric. When G. E. laid off all its carpenters, he left Massachusetts and the United States altogether to return to Canada. But the problem with tramping was there was really no place to go. The Depression was everywhere, as William Ranta of Worcester explained in a letter to the *Carpenter*. " 'Keep out, Keep out!' That is the war cry we hear everywhere. Tens of thousands of our members are hopelessly out of work . . .They are wandering from place to place and when they see those 'keep out' warnings, what should they do?'"[13]

The men who stayed home did what they could. "I would do most anything," says Tom Phalen. "Go downtown, tie on with a trucking outfit for a day. I'd get a buck a day, no matter if it was twelve or fourteen hours, for moving furniture or freight, if they needed an extra." Richard Croteau's father cut ice and bootlegged liquor. Manny Weiner's father picked up odds and ends in home repair work. John MacKinnon sold milk and cream for Morgan Creamery. Enock Peterson and a friend cut wood out in Sherbourne and Holliston to heat their houses and sold the extra for $2 a cord. "Oh God, I'd done everything," says Peterson. "I'd always try to get a few hours of carpentry work. It was enough to keep the wolf away." Oscar Pratt set his carpentry tools aside for two years after he finished his apprenticeship in 1931. Arthur Anctil went for two years without work of any kind. Ed Henley recalls being unable to find dishwashing jobs in restaurants. He had three young children at the time and now claims "if we hadn't taken in boarders, we couldn't have made it."

For a few very men, the hard times of the Depression were someone else's experience. Angelo DeCarlo worked for contractors Sam Abel and Hyman Ecklov. Their firm won enough bids to keep a stable crew. "I didn't lose a day in the thirties," notes DeCarlo. Similarly, Al Valli worked right through the worst years on public construction projects for contractor John Bowen, a political crony of Mayor Curley. The vast majority of capenters had no such luck. Many of them developed job-seeking rituals. These routines helped make the hunt for work more systematic, but just as important, provided some discipline to a life without the built-in structure of an eight-hour work-day. According to Paul Weiner, an alleyway off School Street in downtown Boston became a gathering spot for the city's Jewish carpenters during the Depression. Hundreds of men met there every day hoping, at best, for news of work and getting, at least, company and conversation. MacKinnon developed his own ritualistic method.

I took a dime in the morning and got on the street car and went as far as the street car went into Arlington. Up one street, down another street, looking for jobs. Then I walked back home to Mission Hill. If I didn't get no job, the next day I'd go to Belmont and do the same thing.

A job was a blessing, but not without its own complications. "Everything was cutthroat," says Pratt. On the handful of active construction sites, he recalls, "there were other fellows who were lined up along the fence on the sidewalks ready to take over their jobs the minute they got fired. Things were rough." Every job was precious. The competition for work that always simmered between workers in the insecure construction industry frequently exploded into open warfare. In 1934 a Holyoke business agent refused to allow two Springfield carpenters on a post office job in Holyoke. In reprisal, Leon Manser, secretary of the Springfield Carpenters District Council, threatened to kick every traveling carpenter out of Springfield. In the climate of the Depression years, jurisdictional disputes over as few as two workers quickly evolved into life-and-death struggles. Hoping to avert an extended and bloody combat, Ernest Bessette of the Holyoke CDC set his case before Manser. Bessette's letter reveals how critical every single job was.

90 percent of our membership was out of work and looking for a chance to go to work on this job, along comes two of your members and wants permission to go to work on this job, before any of our own members are put to work, can you think it was possible to sancsion these two men going to work in preference to our own members, just what kind of explanation could the Business Agent have made to our membership if he had done so. Would you have done any different in Springfield.[14]

Long-established working conditions went by the board. After an extended period of unemployment, Harold Rickard was hired on as a carpenter foreman on the Bourne Bridge project. The job operated on a ten-hour basis without overtime pay. Under the circumstances, Rickard says, "no one complained. They went to work." With such a huge surplus of available labor, contractors bore down on the workers relentlessly and without fear. MacKinnon tells a story about Boston contractor John Bowen's working style. "He had the filthiest mouth.

There would be men working down in a hole on a foundation, and he'd be up on the bank cursing and swearing at them. He'd fire five or six and then hire another bunch looking for work, just to get more out of the men."

Bowen's curses were just a slightly more provocative expression of common contractor practices during the Depression. Builders routinely ignored negotiated working conditions and fired union activists who spoke up. Desperate workers reluctantly looked the other way at the growing number of contract violations. For many union carpenters, all the gains they had slowly won seemed to be unraveling before their eyes. There is a sense from letters of the period, newspaper and magazine reports, and even hints in interviews with carpenters fifty years later of a shattering of assumptions, a loss of control, and a feeling of being trapped, of being shaped rather than shaping their own lives. For those who had no work, survival was the first and foremost concern. The few lucky job-holders had other concerns—complicated in a different way. Should they accept a working environment that had been unacceptable a few short years before? If they objected, they did not work. It was, as R. Marlow of Natick Local 847 pointed out, a no-win situation: "No matter where you go you will find many good carpenters walking the streets either through lack of building operations, or because they will not accept employment under the conditions offered by some unscrupulous contractors."[15]

The conditions Marlow referred to included wage payments well below the union scale—as low as the pre–World War I rates of 40 or 50 cents an hour. Union officials tried to slow the practice, but it was too widespread to stop completely. Organizer Charles Kimball publicly urged the expulsion of any union carpenter who accepted below-scale wages. "They are like a cancer, growing from within," he charged, "and are not fit to associate with their brothers in unionism." But even Kimball recognized that such a rigid approach ran the risk of wiping out the bulk of the union membership.[16]

Union officials used the internal system of justice to stem rules infractions. At a September 1932 meeting of the Springfield Carpenters District Council, for example, the business agent brought up thirty-nine union members on charges for working below the union rate. The usual procedures were followed: filing charges, holding hearings, and determining innocence or guilt. But the Depression required more bending than usual in meting out punishment. When three Springfield men were found guilty of lumping, that is, they had installed window casings and laid floors on a fixed sum per apartment basis rather than by the hour, the Council simply directed them to find another job rather than pay a stiff fine.[17] Under depression conditions, a fine was sometimes an impractical method of enforcing discipline. There was just no water to be squeezed out of these rocks.

In early 1930, the Springfield CDC fined Wilfrid Jacques for accepting $1 an hour rather than the prevailing $1.25 rate. Jacques wrote a letter to Recording Secretary Charles Bennett to protest the "very unfair" action. His tale is a typical Depression story and illustrates the tension between the need to maintain union-sanctioned working conditions and the plight of individual working carpenters.

I have been working for this concern for six years and was considered one of their best men. Last winter I had to take a cut in my wages from 1.25 to 1.00 an hour which I fought against as much as I possibly could but the rest of the men wouldn't stand by me. I hated to break the Union rules. I had no choice being hard hit and having a large family to support, but I got our boss to promise he would pay the scale in the early spring which he failed to do. So I went up to him and demanded my full wages which I got, but cost me my job and this $50 fine, while *the rest of the men have benefitted by it and are still working and getting scale wages.*

I think that being a member for seventeen years and never have broken the union rules before must mean something to you. I had to borrow to pay the fine and that has put us in a tough spot. I do hope you will take the depression into consideration and I'm sure you will see my side of the story.[18]

In many cases, the locals did see the men's side of the story. In April of 1930, Charles McIntyre of Local 1105 approached his union to allow him to work below the rate. He told the members that "he found it almost impossible to get a job on account of his age but had a chance to get work if he could get permission to do so." After a brief discussion, McIntyre's request was granted.[19] These isolated union efforts to check or tolerate rampant wage-slashing reminded carpenters of the rules that supposedly governed the job site and shored up faltering union consciousness. But in the final analysis, the impact of all the regulations, charges, reprimands, and fines was swept away by the brutal realities of an economic cataclysm. As long as massive unemployment persisted, employers inexorably drove wages down. Each individual local was powerless to alter the economic context that compelled members to transgress union rules. The ultimate irony was that the locals that fined individual violators were finally forced to accept collectively the very same wage reductions. Had Wilfrid Jacques accepted $1 an hour in 1932 instead of 1930, he would not have been fined. He would have been getting the standard union rate.

In the summer of 1930, the *Builders' Record* reported that union craftsmen were "offering" to work for less than the union scale. Therefore, the BTEA publication argued, Boston's employers should be released from their contractual obligations with the unions and allowed to lower their employees' pay. The surprised Building Trades Council, accustomed to several years of cordial relations with the BTEA, called on BTEA Secretary John Walsh to "substantiate with facts his unsupported statements."[20] In reality, both sides were shadow-boxing, dancing around a new and potentially dangerous form of negotiations for the unions. The contractors not only wanted reductions, they wanted them in the midst of the term of a collectively bargained agreement.

Other than the open-shop disaster of 1921–22, Boston's union carpenters had never taken a wage reduction. Even then, the final agreement maintained the old wage rate. All along, the unions had refused to officially accept the 10 percent cut that prevailed during that eighteen-month strike. In addition, the city's carpenters had never opened an existing settlement for employer concessions. Now, the BTEA wanted both. It was an unprecedented request, but both employers and workers recognized that the severity of the Depression was unprecedented as well. Carpenters may not have been "offering" to work for less, but they were certainly accepting less. The contractors were intent on using the drop in building activity as a lever to lower labor costs substantially. Workers prayed that the Depression would be short lived. As the unions saw it, their best chance was to tread water and hope that construction demand would rebound in time to avoid too many serious losses.

The initial flurry of charge and counter charge brought no immediate action, but the issue had been placed at the top of the negotiating agenda. In October of 1931, Robert Whidden, vice-president of the BTEA, contacted all the building trades unions in the city. Calling his letter "simply a friendly overture," Whidden argued that construction recovery was impossible at the existing union pay scale. With wage reductions, however, he assured the unions that Boston's builders and their construction clients would find the incentive to gear up for renewed production. A month later, John

Walsh announced that the BTEA was not looking for a "permanent reduction in wages, but rather an earnest desire to put forth a measure of mutual helpfulness forced by extremely bad business conditions."[21]

Again, the unions [along with Mayor Curley] rejected the BTEA proposal. But without economic recovery, the cuts could be staved off only so long. The crisis showed no signs of easing. On the contrary, it seemed to many that a permanent state of depression had settled on the population. "I have a definite feeling," Louis Adamic said of American workers in the dreary days of 1931, "that millions of them, now that they are unemployed, are licked." BTC officials refused to accept Adamic's pessimistic description. No one understood the extent of the Depression better than construction workers. But that did not mean they had to act as if they were licked. Boston union leaders did not accept the logic of the employer prescription of reduced wages as a cure for their ills. The situation was bad, even desperate. But in the minds of the BTC officers, there was no proof that wage concessions would result in anything other than a fatally weakened trade union movement. The UBCJA national office sent troubleshooter T. M. Guerin to Boston once again to resolve the impasse. Guerin cajoled, wheedled, and twisted arms. By late November, he managed to convince local Carpenters officers and enough other construction union leaders to accept a 20-cent-an-hour reduction across the board. Starting January 1, 1932, BTEA members paid union carpenters $1.17½ instead of $1.37½ an hour.[22]

The concessions in Boston triggered cuts in other cities. In the summer of 1931, delegates to the Springfield District Council had unanimously approved a resolution stating that wage cuts would not "be of any benefit to us." The following spring, four months after the new Boston rate went into effect, Springfield builders reduced their carpenters' paychecks by $2 a day without serious opposition. Occasionally,

the concession fever did not even require employer initiative. On February 4, 1932, the Greenfield local voted to lower their wages from $1 to 90 cents. There had been no employer demand or hint of a demand. The Greenfield carpenters assumed it was just a matter of time and decided to offer a gesture of good faith by voluntarily suggesting the reduction. Town after town followed the concession pattern. When the dust finally settled, Boston's carpenters remained the highest paid in the state. Outside the capital, no union carpenter earned more than $1 an hour.[23]

The draconian cuts of 1932 whetted the contractors' appetites. They wanted more and they wanted it quickly. Twelve months later, the employers in seven Boston craft associations proposed further reductions ranging from 25 to 45 cents an hour. But the Depression had bottomed out by the winter of 1932–33. Worker suffering was so widespread that it become harder for employers to justify and win support for even greater sacrifices. In February, a joint meeting of the Building Trades Council and all the other construction unions outside the Council (including the Carpenters) agreed to develop a united front against the employer demands. Representatives of the meeting notified the BTEA that lowered pay in any single trade would immediately produce a city-wide general construction strike. The threat worked. The cuts were never implemented. The union carpenters' scale in Boston did not change until they won a modest raise in 1936.[24]

Again, developments in Boston influenced other cities. In May 1933, the Master Builders Association for Springfield, Holyoke, and Chicopee suggested another 25-cent reduction for area carpenters. As in Boston, the Springfield and Chicopee builders backed off in the face of a unanimous strike vote. Holyoke and South Hadley contractors, on the other hand, went ahead and managed to run jobs at 75 cents an hour for almost two months. After a strike threat and extended negotiations, the builders eventually relented and returned to the previous rate.[25]

The beleaguered unions were clawing to survive. Fighting off plummeting wages and disappearing work rules on the job site, the locals also faced internal difficulties. The main problem was financial. Without any income or job possibilities, many members dropped out of the union. It was not a question of diminished union sympathies. It was simply, as John Greenland notes, that they "couldn't afford to pay dues." The unions desperately sought to hold the members together. Monthly dues were lowered and the national UBCJA office ruled that members were allowed to be behind in payments for a year without facing suspension. In 1932, over one hundred thousand union carpenters across the country were in arrears from three to twelve months. The Worcester Carpenters District Council imposed a $1 a day assessment on working members to assist others who were unemployed.[26] Individual carpenters lucky enough to have jobs took it upon themselves to help out less fortunate brothers. Enock Peterson covered dues payments for dozens of unemployed members. Al Valli took $150 out of his personal bank account to pay his local's business agent when the union treasury ran dry. Harold Rickard hired as many carpenters as he could get away with when he was foreman on the Bourne Bridge.

All these charitable acts of assistance were hopelessly inadequate. More and more locals operated in the red. The carpenters in the shoe town of Newburyport had seen their local go broke back in 1925. The Newburyport experience was increasingly repeated as the depression became universal. Between 1929 and 1935, two UBC locals in Gardner, and one apiece in Canton, Ipswich, Shrewsbury, Taunton, Webster, and Wilmington either consolidated, let their charters lapse, or disbanded completely.[27] The less marginal locals survived but with far fewer numbers. Between 1930 and

1933, the combined building trades unions lost over three hundred thousand members nationwide. The Massachusetts carpenters locals were no different. (See the table below.) The 1933 state convention was canceled because too few locals could afford to send delegates.

Town	Local	Membership		
		1917	1927	1937
Boston	33	659	1,054	459
Lawrence	111	292	212	104
Lowell	49	287	216	60
Springfield	177	529	359	236
Taunton	1035	113	96	43
Worcester	408	211	154	51

Source: UBCJA membership records, national office.

The locals devised a variety of strategies to protect the members and keep the organizations afloat. Greenfield Local 549 tried to refuse any new entries in order to preserve the limited job opportunities for existing members until the national office overruled their protectionist action. Other locals rolled back officer salaries or turned the post of business agent into a volunteer position staffed on a rotating basis by jobless members. As the Depression dragged on, a growing number of carpenters raised the notion of work-sharing as a means of equalizing distress. At first, most unions resisted. The very idea cut against the grain of the carpenters' work ethic. There was no doubt that building tradesmen eagerly sought collective solutions to counter employer domination. Over the years, they had created an elaborate set of rules to govern the workplace and a democratic system of justice within their unions. They willingly abided by a standard wage and a collectively bargained agreement. But within that cooperatively developed framework, it was an unassailable point of pride that success in the industry was a matter of individual achievement. If one carpenter worked more steadily than another, presumably that status was a testimony to greater competence, skill, and perseverance. Carpenters knew that stable employment with one firm was not always based on merit since favoritism and nepotism were ever present. Regardless, the hiring, firing, and promotion of individual carpenters had *never* been an arena of union intervention unless it was directly related to union activity. Voluntary assistance by a working carpenter to an unemployed member was entirely consistent with the unions' culture of solidarity. Advocating even minimal mandatory allocation of jobs, however, was a daring and collectivist challenge to a traditionally highly individualistic domain.

The unions eased uncomfortably in this direction. By 1929, many locals required union-granted permits on overtime jobs. Fines were levied against members who worked more than eight hours without permission. Overtime bans, however, just scratched the surface of the job shortage crisis. In 1931, a delegate to the Springfield Carpenters District Council proposed a far-ranging twofold approach. He called for a hiring list that ordered preference on the basis of length of unemployment. Its universal application would be enforced by a fine on any member who failed to register. In addition, the delegate proposed a system of staggered work crews in which every contractor would be required to hire one group of carpenters one week and another crew the next week. The proposal proved premature. Opposition was so strong that even the author backed down. The resolution was defeated 50 to 0.[28]

The unremittingly high unemployment figures persisted, and soon once radical-sounding suggestions appeared increasingly reasonable. With time, Springfield's carpenters reconsidered their opposition to hiring restrictions. In February of 1932, the District Council endorsed the staggered system of two weeks on and two weeks off for crews on city work. The principle still grated, but "we feel," said the Council, "that it is the best that can be got at the present time." Carpenters in neighboring

Holyoke also adopted the staggered system until the town goverment abused the mechanism. In March of 1932, the Holyoke Board of Public Works decided on a policy of open shop jobs with a union crew one week and a nonunion crew another. At the request of the Painters' Union, the Holyoke Building Trades Council placed the Board on its unfair list. With jobs so few and far between, the Holyoke Carpenters District Council supported the BTC action with great reluctance. They issued a statement attacking the city's antiunionism, but opined that "the so called stagger system . . . is the best that can be got from the present administration."[29]

The unions tried a host of other defensive measures. The Springfield CDC urged contractor E. J. Pinney to institute a six-hour day on his Technical High School project. In 1933, the Boston building trades unanimously endorsed an emergency 24-hour work week (three eight-hour days) and a negotiated thirty-hour week (five six-hour days) in the next contract. The Worcester Central Labor Union opened a campaign against the use of labor-saving equipment in construction after they learned that the four major city projects totaling $2 million employed only fourteen carpenters. Springfield's carpenters aggressively lobbied local politicians to support a concrete (rather than steel) design for the proposed Ludlow-Springfield bridge over the Chicopee River.[30]

In the long run, the defense of the union wage, the assessments on working carpenters, the declining dues structures, the shorter days, and the staggered crew system, barely made a dent in the overwhelming unemployment problem. Sharing the available work was admirable and highly principled, but more work was the only long-term solution. Jobs were what was needed, and the longer the Depression wore on, the clearer it became that the economy would and could not miraculously right itself. Employer promises of a building recovery in the wake of pay cuts had turned sour. Though few unionists ever accepted that argument, even the slimmest of hopes die hard. Construction activity continued at a snail's pace and showed no signs of escalating. If conventional private-sector construction had reached a state of permanent collapse, where could building tradesmen turn for the promise of better days?

12

Work, Not Relief

The election of Franklin Delano Roosevelt in 1932 carried with it a vaguely defined but undeniable shift in public expectations of the federal government. In the grim months following the stock market crash of 1929, President Herbert Hoover had repeatedly refused to acknowledge the extent of the crisis. He feared gloomy public statements from the White House would only further undermine business confidence and, instead, was content to restate his long-standing opposition to large-scale federal intervention. "We cannot," he told a press conference in 1931, "legislate ourselves out of a world economic depression."[1]

In the absence of any positive economic markers, Hoover's stubborn passivity brought him little more than ridicule and contempt. By the time of the 1932 presidential campaign, a majority of the electorate had lost patience with political nostrums that continued to sanctify the wisdom of the business community. The doctrine of unfettered capitalism had been discredited. The promise of permanent prosperity, so widely hailed in the "New Era" of the 1920s, seemed like a cruel joke in 1932. In just a few years, the American businessman had been stripped of his statesmanlike aura

and singled out as the primary villain of the Depression. Even Herbert Hoover had been known to join in the chorus of catcalls. "The only trouble with capitalism is capitalists," he once commented. "They're too damned greedy."[2]

The election results signaled the voters' desire for a new and different role for the government. Despite Roosevelt's cautious campaign rhetoric, the symbolic contrast with Hoover was clear. A victory for the Democrats in 1932 represented an endorsement of a more activist federal government, particularly in the area of relief and job creation programs. It was a mandate that the new administration took seriously. In the first hundred days of office, FDR and his Brains Trust took steps to intervene in banking, farming, home mortgages, land use, and economic planning. And, most important for the nation's building trades workers, the New Deal inaugurated an alphabet soup of agencies (PWA, CWA, CCC, and WPA) to revive the moribund construction industry.

The frenetic pace and broadened scope of federal activity unnerved much of the national leadership of the AFL and its affiliated unions. The first thirty years of the twentieth century had cemented a mistrust of government agencies among Federation chieftains. Having been victimized by antitrust legislation, court injunctions, and federal troops, many labor leaders feared the repressive power of the state and, accordingly, adopted a hands-off attitude. Samuel Gompers had argued, at times, that individual liberty and an interventionist state were polar opposites, that one could only be achieved at the expense of the other. As a result, the AFL had generally opposed reform legislation that required state or federal regulation. Testifying against a proposed social insurance bill, Gompers once explained:

Sore and sad as I am by the illness, the killing, the maiming of so many of my fellow workers, I would rather see that go on for years and years, minimized and mitigated by the organized labor movement, than give up one jot of the freedom of the workers to strive and struggle for their own emancipation through their own efforts.[3]

Gompers's death produced no shift in the political alignments of organized labor. William Green subscribed to a similar voluntarist ideology. The AFL's views had meshed neatly with the laissez-faire political climate of the twenties, but Roosevelt's ascendancy forced the Federation to come to terms with the overwhelming national sentiment for major social change initiatives out of Washington. Perhaps no single AFL affiliate was less ideally situated to accommodate this new public mood than the national UBCJA. Not only was the leadership of the Brotherhood wary of FDR's reform proposals, they were one of the few groups of labor leaders to endorse and work for Herbert Hoover in 1932. General President Hutcheson managed the labor bureau of the Republican party during the campaign.

William Hutcheson was a life-long Republican. At various times, his name had been floated as a possible candidate for secretary of labor and even vice-president in a GOP administration. Hutcheson had indelibly impressed his personal stamp on the UBCJA ever since he took the helm in 1915. He brooked no dissent at the national level. Brotherhood officers and staff carefully toed the political line he advocated. As a result, the national union was a solid bastion of Republicanism in a predominantly Democratic, if conservative, AFL. When the Federation departed from its customary distaste for independent political activity and endorsed the presidential candidacy of Robert La Follette of the Farmer–Labor party in 1924, the Brotherhood was the only major AFL affiliate that refused to participate.

Roosevelt's job creation programs put the union in an awkward position. Hutcheson's extreme version of AFL voluntarism could not tolerate the notion of the federal government as an employer of last resort. He was happiest when the government operated at a long arm's length from the labor movement. He preferred to deal with management directly without the mediating influences of state and federal agencies. In part this was a tactical judgment. Compared to the limited power of decentralized building employers, the authority of the federal government could easily abridge the significant influence of the building trades unions. That was why Hutcheson had adopted such an uncharacteristic posture of militance during World War I. While other labor leaders viewed wartime federal guidelines as an opportunity to expand the ranks of organized labor, Hutcheson primarily saw the dangers, both immediate and potential, of a voracious federal appetite eating away at collectively bargained agreements. Hutcheson never dropped his passionate distrust of state intervention, consistently preferring the known quantity of direct labor–management negotiations. In 1937, he unsuccessfully opposed AFL endorsement of minimum wage legislation. A full decade after the onset of the Great Depression, Hutcheson remained one of the few public figures who still opposed New Deal legislation and the accompanying expanded role of the state. Labor "has known that what government gives, government can take away," Hutcheson said in a 1940 speech. "Where government has failed, labor and industry can succeed . . . in spite of every handicap that government may place in the way."[4]

Early in his term of office, Hutcheson eradicated whatever vestiges of labor radicalism persisted from the McGuire era at the Brotherhood's national level. The general president crushed any and all opposition. But though he was never forced to yield on his machine rule or his political conservatism, the human wreckage of the Depression compelled Hutcheson to grudgingly reconsider his repudiation of all federal jobs policies. The rank-and-file carpenter did not have the security of a union office and salary to cushion his suffering. He yearned for relief of any kind and cared little about the source. Many carpenters shared Hutcheson's misgivings about an expanded federal presence in construction but few had the option to turn aside publicly funded employment in order to maintain ideological purity. With their reluctance to advocate unemployment relief and public works programs, Hutcheson and other comfortable AFL leaders showed themselves to be out of touch with the needs and perceptions of the average union member.

Like most working-class Americans, the majority of building tradesmen had lost faith in the ability of the private sector to generate sufficient jobs and welcomed New Deal efforts to provide supplemental employment. Massachusetts carpenters embraced the Roosevelt administration. "When FDR took over the reins, things began to improve," John MacKinnon says. "People got back to work. It was a lot better. I give FDR all the credit in the world." Enock Peterson praises Roosevelt in similar glowing terms. He worked for the WPA, building schools, grandstands, and athletic fields.

The New Deal saved him, Peterson claims. "If it hadn't been for Roosevelt . . . ."

Local union officers lived with the Depression more intimately and, as a result, understood the distress of the membership more acutely. They did not necessarily share the voluntarist principles that hamstrung national AFL leaders. On the contrary, many local building trades officials in Massachusetts led the charge for political reform agendas and pro-labor candidates. The Boston Building Trades Council, for example, called for an independent Labor party in 1932, a position considered heretical by top AFL officials. The Council actively pressured municipal, state, and federal agencies to seek out either private or public loans to subsidize construction projects. Over a thousand carpenters and other craftsmen rallied at the Parkman Bandstand on the Boston Common for a June 1933 BTC-sponsored "work demand" meeting. Massachusetts AFL President James Moriarty told the assembled construction workers: "The key to our whole difficulty is to put people to work. Work instead of relief means that millions can buy bread and retain their self-respect . . . we shall not be content with continued pauperism and doles."[5]

The distinction between work and relief was a constant theme for Depression-era carpenters. They wanted work. Relief may have been unavoidable after extended bouts of unemployment but it never became desirable. In the exclusively male world of construction, the equation of relief and dependence was particularly powerful. The rigidly defined sex roles of the era reinforced carpenter heads-of-households' discomfort with even token assistance. The very thought of charity was humiliating for men who prided themselves on their fierce independence. They had learned how to cope with a chronically insecure occupation without resorting to private aid or public dole. They had overcome the lack of stability with a combination of individual mastery of their craft and collective control of the job site through their unions. They had used the tools of the trade to raise and support families. Now all those years of effort were unraveling uncontrollably. The deft touch of the master carpenter was useless without a job, and the carefully constructed work rules went by the board when union members were forced to accept substandard wages and working conditions.

The unemployed often blamed themselves for their plight. The common wisdom of the 1920s linked workers' modest prosperity to their own hard work. The cultural pundits of the New Era dismissed larger social and economic forces and attributed the twists and turns of personal histories to individual choices. When conditions changed, it was almost inevitable, as historian Robert McElvaine points out, that "Americans who had claimed responsibility for personal gains found it difficult not to feel guilty when confronted with failure."[6] Carpenters who accepted the notion that the triumph of an adequately furnished home and a slim nest egg of savings rested solely on their shoulders were at a loss when the Depression hit. They were willing to work and push themselves just as hard, but somehow, that was no longer enough. In the early years of the crisis, before its full impact was completely understood, many of the unemployed assumed they were at fault for their own situation.

The idea of unemployed relief assistance, therefore, ran counter to most of the assumptions and aspirations of American carpenters in the early 1930s—including those who were out of work. The national UBCJA rejected the concept as an illegitimate federal policy, and rank-and-file carpenters viewed relief as a stigma, a stinging brand of personal failure. To accept relief with equanimity, a jobless worker must feel "entitled" to it. The feeling of entitlement depends on a recognition that one's personal

misfortune is, in large part, due to forces out of one's control. Attitudes toward relief shifted dramatically in the mid and late 1930s. By that time, most American workers had come to the conclusion that their suffering was caused, not by their own decisions, but by national and international economic developments. Still, the initial horrors of the dole never completely disappeared. Thus, whatever other differences existed, the absolute political priority of jobs creation programs over a more comprehensive relief system united every Brotherhood member, from Hutcheson to the unemployed carpenter on the street.

Union carpenters welcomed Roosevelt's stabs at construction works programs but, like Hutcheson, they ultimately preferred private building. The memories of World War I federal meddling with the closed shop were only fifteen years old. But the mistrust ran deeper. The industry had evolved its own set of idiosyncrasies that building trades workers liked to think made them unique and incomprehensible to outsiders. Even with a legacy of bitter employer–employee conflicts, tradesmen were far more likely to consider contractors, especially those who came up from the ranks, as part of the "club" than any state or federal bureaucrat. Unlike a factory owner, the head of a construction firm was usually visible to the workers, occasionally exchanging pleasantries on a first-name basis. The builder knew the industry, the tools, the terminology, and the feel of a construction site. He was a member, if an unequal one, of the construction family and, most important, he had a tradition of collective bargaining. Managers of public works agencies shared none of those bonds. They were government appointees whose loyalties to the political party in power transcended any affinities for the industry.

Initial public policies did nothing to overcome the misgivings and allay fears of a replay of the World War I experience. Government officials were seeking methods to put people back to work. Carpenters acknowledged that work had to be found for all the unemployed, but they did not want their industry to become the repository for every jobless American. Having spent their working lives in the construction field, carpenters expected that construction jobs programs should give priority to experienced craftsmen. In addition, they believed those jobs should be carried out under customary, i.e., union, working conditions. They found, to their dismay, that municipal, state, and federal officials did not necessarily agree. On the contrary, some politicians used the opportunity to vent their antilabor animus. In the winter of 1931–32, for example, the Worcester Mayor's Unemployment Committee referred unemployed tradesmen to the notoriously antiunion E. J. Cross at 50 cents an hour.[7]

The first federal and state projects also ignored labor agreements. Ovila Marceau worked on a Federal Emergency Relief Administration project in western Massachusetts for 50 cents an hour. In 1933, when the National Recovery Administration funded several highway construction contracts in Massachusetts, the state Commission on Public Works set the pay scale for all skilled workers at 55 cents. The following year, the City of Springfield paid carpenters from 56 to 64 cents for concrete work on the city sewer system. The UBC resented the sabotage of the union rate with taxpayers' money. The Massachusetts State Council of Carpenters initiated a statewide letter campaign to Governor Joseph Ely to protest the 55-cent highway rate, charging that it was "a sweat shop rate and in direct violation of the spirit and intent of the N.R.A."[8]

Despite the low wages, publicly funded projects offered the only alternative to unemployment. In 1933, the Springfield District Council granted carpenters on welfare permission to do carpentry work at the city farm as long as the welfare board gave them $1 credit for every

hour they put in. Carpenters were careful to distinguish between criticism of the programs and of their abuses. A January 10, 1934, resolution from the State Council wholeheartedly supported the NRA, Public Works Administration (PWA), and Civil Works Administration (CWA) but also noted that "some of the officials in the several cities and towns who are handling the C.W.A. programs are not paying the stipulated wages and some of the contractors on P.W.A. are violating the intent and meaning of the act as it relates to hours and wages." As a solution, the Council suggested uniform statewide standards for all the programs, including a $1.20-an-hour wage.[9]

For the next few years, building trades unions and New Deal administrators haggled over wages and hiring policies. The union position was straightforward—preferential hiring for union craftsmen, union pay scales, and publicly funded private contracts rather than direct government employment whenever possible. Roosevelt's position was less clear. Administration officials constantly developed and redeveloped regulations as Congress allocated more funds and created new agencies. Through 1935 and 1936, guidelines on PWA and Works Progress Administration (WPA) projects shifted from month to month depending on the individual state administrator, the political influence of the unions, and the latest directive from Washington.

In July of 1935, the Massachusetts Building Trades Council met to consider a proposed six-state strike against a suggested PWA and WPA hourly scale of 65 cents for 130 hours of work a month for all skilled mechanics. Throughout the summer and fall, BTC leaders negotiated with Arthur Rotch, WPA administrator in Massachusetts. By the beginning of winter, Rotch received an order from Washington allowing him to set wages at the prevailing rate. However, as soon as the unions had won the battle on wages, Harry Hopkins issued a series of contradictory hiring directives from

Washington that reignited union anger. Hopkins, FDR's right-hand man, had agreed with many of the unions' arguments when he accepted the top post at the WPA. Though his primary responsibility was to provide as many jobs as possible within budgetary constraints, Hopkins announced that he would not accomplish his mission at the expense of the private sector and its system of collective bargaining in the construction industry.[10]

In August 1935, Hopkins issued Administrative Order No. 15 requiring WPA projects using private employers to give union workers preference. This directive simply maintained traditional employment practices in the industry. Four months later, Hopkins amended the order. He still allowed union preference but only after priority had been given to residents of the "political subdivision" in which the work was being carried out. This effectively knocked out any union tradesman who lived outside the local project area. Once again, the Massachusetts BTC threatened to strike. In early January 1936, the WPA head countermanded both of his previous orders and ruled that 90 percent of all WPA project employees must be drawn from the relief rolls of the U.S. Employment Service. The BTC immediately moved from threats to action. Six hundred craftsmen in Boston and Newton and two hundred more in Waltham, Concord, Framingham, and Natick walked off WPA jobs in protest.[11]

Hopkins's ruling angered union workers. The loss of the protection of union status was bad enough. It appeared, in that sense, to be a rerun of the Gompers–Baker pact. In addition, they believed that the quality of construction would inevitably suffer as untrained workers attempted to carry out skilled tasks. But the underlying assumptions behind preferential treatment for those on relief irritated union tradesmen the most. As they saw it, the WPA would reward those who chose the dependent life of the dole and penalize those who had been too proud to accept charity. For the next two

months, a comedy of bureaucratic bungling characterized the dispute. The WPA rescinded the ruling on January 18, only to restore it on February 25. Finally, on March 16, Massachusetts WPA director Paul Edwards agreed to a 50–50 split between the relief rolls and the union hiring halls. *Our World*, a Boston labor paper, reported satisfaction among building trades leaders with the settlement. "This means that those men whom the depression had hit as hard as any others but yet had preferred suffering rather than stoop to beg relief, will no longer be denied jobs as union men who know their job, irrespective whether their names are on the relief rolls."[12]

In Massachusetts, at least, organized building trades workers had resolved their basic differences with the WPA. Periodic flare-ups continued to occur, but the unions generally refocused on private contractors, such as John Bowen, who persisted in paying below the rate or hiring nonunion tradesmen. The employment situation improved slightly in the second half of the decade though it never dropped into single digits. Raises in 1936 and 1937 brought the union carpenter's wages back to the pre-Depression high of $1.37½. The economy had by no means reached a state of normalcy, but mild gains combined with the dramatic upsurge of organizing among industrial workers encouraged construction workers to return to some of their more militant habits. In September 1937, for example, 330 carpenters and other tradesmen walked off the $3.7 million seventeen-story Suffolk County Courthouse to protest the presence of nonunion granite cutters.[13] A sympathetic action (or a construction project, for that matter) of that scale would have been unthinkable five years earlier.

Despite the mild rebound in the private sector, the dominant themes for union carpenters in the 1930s were wrapped up with their ambivalent feelings toward the New Deal and the federal role in the construction industry. There was no question that the average union carpenter identified with and embraced New Deal initiatives to relieve unemployment. It was equally clear that union leaders and members never abandoned their qualms about the government as employer. History had taught them a healthy respect for the ability of the state to undermine labor relations. The early practices of the FERA, NRA, CWA, PWA, and WPA had only confirmed these fears.

But it was the question of relief and public works that encapsulated all the complex political strains. Union officers attempted to resolve their conflicting attitudes by viewing the federal government as just another employer. By and large, they were unwilling to enter discussions about the proper role of the state in social and economic policy. They preferred a concrete hands-on approach. That is, if the government planned to act as a major consumer of construction services, the unions insisted that it play by the same rules as any other employer. Most construction union leaders had little use for Keynesian pump-priming theories. They rejected any notion of substandard wages as a method of sharing the pain or kicking off economic recovery. If the government wants to build, union leaders proclaimed, it must operate under guidelines established by collective bargaining. Any other approach or consideration could only undercut decades of union struggle. It was as simple as that.

Union solutions were equally simple. Most union officials welcomed federal funding, but all hoped to minimize direct government administration. They wanted to avoid any direct political control over construction, preferring the familiar adversarial relationship with contractors and employer associations. In a lengthy resolution passed unanimously at the 1936 UBCJA national convention, the delegates expressed their satisfaction at the prounion reforms instituted by the WPA but, on balance, maintained a highly critical position. They advocated taking construction management decisions out of the hands of federal and state administrators and turning "this work over to *our* [my emphasis] general contractors for supervision, contractors who are equipped to do this work more efficiently." Should that plan fail, continued the delegates, "we ask that [WPA administrators] place on all future work, union skilled mechanics as foremen and supervisors to whom it rightfully belongs." Interestingly, the hostility to government administration was so strong that conventional private construction management was deemed superior to weak and inefficient federal administrators dependent on the craft knowledge of union workers. The perspective embodied in this resolution cannot just be chalked up to Hutcheson's fanatical voluntarism. At the 1938 Massachusetts AFL convention, all forty-two building trades delegates introduced a similar resolution. They charged that the WPA was competing with the "normal" construction industry and, as a result, "seriously retarding recovery in that industry." They proposed that any project over $10,000 be put out for bids by private contractors rather than be administered by government agencies.[14]

Rank-and-file carpenters rarely voiced a purely ideological opposition to public intervention. They shared their leaders' distrust of governmental intentions regarding wages and working conditions. But they also wanted to work and they knew, as Springfield Business Agent Harry Hogan told a "somewhat startled" Central Labor Union meeting in 1939, that "carpenters can do better on the WPA than in private employment." They supported the New Deal's political programs. For the rank-and-file carpenter, the controversial issues of relief and public works were more personal. The 90 percent rule had provoked intense hostility. The administration order had tapped into something very deep inside those workers. Why else walk off a job when hardly any jobs are available?

Carpenters had constructed working lives based on an ethical system revolving around the twin issues of independence and cooperation. The independence, or self-reliance, served a crucial function. It helped fashion an unshakable work identity that transcended any particular contractor or any particular project. The carpenter's identity was based on his tools, his knowledge, and his own efforts. He stood alone, making a living by his wits, employed by others, but not defined by any single employer. Yet that independence was tempered by an equally strong sense of cooperation. The decades of labor wars in construction had demonstrated the necessity of collective action to even the most individualistic of carpenters. Their very livelihoods, their pursuit of independence in their careers, ultimately depended on their ability to cooperate successfully in the form of trade unionism. Thus, Harry Hopkins's 90 percent rule was, in their minds, a slap in the face. They felt entitled to work, not to relief. Preferential treatment of those on relief told these carpenters that their values and choices would bring no rewards at the most stressful point in their working lives. It was a message from their government—their democratically elected representatives—that dependence and defeatism paid off.

They were, as one tradesman put it, "self-sustaining" workers, and their sense of self-respect depended heavily on that self-image. The Depression had wrought serious damage to that image. It was more than just a question of where the next meal would come from. Preoccupation with economic survival invariably triggered corollary concerns—family crises, alcoholism, depression. Carpenters had sustained themselves through previous difficult periods with the help of their own culture of cooperation and independence—a culture peculiar to the building industry and nourished by its unions. The battles over relief and public works were clearly matters of bread and butter. But they were also symbolic struggles over the acceptance or rejection of the value of the carpenters' culture and his sense of self-worth. The reinstatement of the prevailing wage and the defeat of the WPA 90 percent rule brought union carpenters more and better-paid employment. Those rulings also validated a lifetime of personal choices. As a Boston labor reporter wrote, building trades workers "stressed that labor's dignity demands jobs be given in labor's own name without a thought of charity."[15]

13

Jobs, Jobs, and More Jobs

New Deal legislation created jobs for millions of Americans, but it did not bring recovery. In fact, the Depression outlasted the New Deal. Under pressure from an increasingly conservative Congress, Roosevelt had backed away from his more ambitious programs by the end of the 1930s, steering a cautious and conciliatory course. Administration policies had helped to lower the unemployment rate to a limited extent—from the high twenties to the teens. But in 1939, almost 9.5 million Americans were still out of work. It took four more years to reach pre-Depression levels. The return to full employment had, in the final analysis, little to do with the reform programs of FDR's "Brains Trust." Only the monumental task of preparing for entry into World War II was finally able to generate enough work to erase the suffering of the jobless.

The build-up began in 1939 and instantly restructured economic priorities. Accustomed to more than a decade of lean years, Massachusetts carpenters suddenly found themselves in the midst of another war-induced construction boom. "All the private work died," recalls Paul Weiner, and so did much of the

civilian public work. The *Boston Herald* reported that by October 1940 many WPA projects in the commonwealth had to be suspended because so many skilled craftsmen had been diverted to military construction.[1] Carpenters were set to work on the Boston Army Base, Air Force bases in Weymouth and Westover, the extensive expansion of Fort Devens, and the mammoth $29-million military encampment project at Camp Edwards.

All the work was welcome, perhaps none more so than that at Camp Edwards. For the first time since the early 1920s, carpenters in troubled southeastern Massachusetts found a stable source of employment. Men traveled from the Fall River, New Bedford, and even Brockton locals across the Cape Cod Canal to build the camp. Joseph Leitao got hired on—his first job as a carpenter since leaving his native Portugal in 1919. Though Leitao had been shut out of his trade for twenty years, he had retained his skills. As American involvement in the war stepped up after Pearl Harbor, contractors were delighted to find trained men like Leitao. The labor pool was shrinking as younger carpenters signed up for the armed forces

130

while the demand for their labor was expanding.

The military was "crying" for skilled workers, Leo Bernique says, and, by and large, did not find them. Manny Weiner's father worked on a military base alongside tailors and shoemakers. Anyone who had the nerve to say he was a carpenter got a job, says Weiner. Camp Edwards provided a haven for the legions of unemployed shoemakers from Brockton, according to Oscar Pratt. Fort Devens served the same function for the unemployed of northern and central Massachusetts. Thousands of workers labored seven days a week at Devens. "We had butchers, barbers, bakers, everything," chortles Thomas Phalen. Two or three hundred people at a time went to the Devens employment office. "If you had a union book," says Phalen, "all you had to do was hold it up and you'd get a job. Every foreman wanted you because it made things easy for him." Military needs overrode all other considerations. Most jobs worked the men as long as they could stand it. "Quite a few of us went down to Castle Island in South Boston," remembers Ellis Blomquist. "We were there for six or seven months shoring cargo. We were working two days and three nights right through just to get the ships out. You didn't dare go near a wall, because you'd fall sound asleep leaning against it."

Since the military administrators of World War II respected local working conditions, the work was unionized and the new "carpenters" joined the Brotherhood. They were not always received with open arms. The construction relieved some, but not all, of the anxieties of long-standing members. No one knew how long the war would last nor what to expect after the war. The memories of the Depression were too recent and vivid to welcome comfortably the potential competition of hundreds of new members in each local. Furthermore, the new workers had no sense of union tradition and little motivation to develop one. They rarely

attended union meetings and, says Pratt, "had no concept of what a union was all about. They just knew that they had to join a union to get a job."

The great majority of the wartime tradesmen later returned to their previous occupations. "You had to have them during the war," nods Phalen, "and they'd use them wherever they could—boarding up, nailing, shingling. But when the job was over, they just disappeared." Still, their sheer numbers had an impact on the unions in the war years. Membership in the Fitchburg and Leominster area locals spurted to eight to ten thousand. A number of locals took advantage of the new population to replenish their depression-ravaged treasuries. Worcester, Lawrence, or other outside carpenters seeking work at Fort Devens were forced to pay an extra initiation fee to the local unions. New Bedford's Local 1416 refused to go along with the common union practice of lowering initiation fees for the Camp Edwards project. Sticking to the standard $75 price of admission, Local 1416 rebuilt its bank account from $1,000 to $262,000 in one year. A *Herald* reporter covering the story claimed that other New Bedford unions were "down on the carpenters" for their policy.[2] Given the overwhelming public support for universal sacrifice for the war effort as well as the already strained relations between prewar and wartime carpenters, the decision of several locals to profit from the new men's presence inevitably created, as Paul Weiner indicates, "a lot of hard feelings."

The government came in for its share of hard feelings. In 1942, the Roosevelt-appointed War Labor Board developed a set of guidelines for wartime labor relations. Under the banner of "equality of sacrifice," the Board offered union shops to organized labor in exchange for wage freezes, compulsory arbitration, and a no-strike pledge. The trade-off appealed to many workers who had been struggling to create or stabilize labor organizations, particularly in industrial settings. For union carpenters with a long history of a closed shop, the sacrifice was obvious and the benefits nonexistent. The Board's policy provoked scattered wildcat strikes and, ultimately, a major confrontation with the United Mine Workers. The Carpenters tolerated the ruling. The lack of organized resistance, however, did not imply contentment, says Tom Harrington. It was "sand in the claw for myself, my father, and all union men that I knew, that their wages were frozen but the profits of the contractors were never frozen. It seemed like the rich got richer and the poor stayed just the way they were." Harrington and others waited restlessly for the end of the war to bring their brothers back from overseas and to correct inequities at home.

———

The veterans who returned from the European and Pacific fronts found a country ready to build. The hardships of the Depression and the imposed sacrifices of the war years were over. American industry was preparing for a shift to peacetime production and the American people were ready to spend money for long-delayed consumer necessities—clothes, appliances, cars, and above all, houses.

"After the war, there were just not enough houses," recalls Ernest Landry. He went to work on a 475-unit project in Chelsea. Chelsea was not alone; in every town across the country 2 × 4 wood frames sprouted like weeds, on individual lots or in the instant communities of housing developments. From 1945 to 1955, private housing went from a $1-billion to a $22-billion industry. The ballooning economy required nonresidential work as well—new plants, warehouses, and offices. Nationwide, commercial construction expanded by 1,485 percent in the decade following the war.[3] The volume of the work was unprecedented, as was the scale of each job. Contractors continued the pattern set during the war, hiring a hundred or more carpenters per site. Richard Croteau

remembers the period as "unbelievable." All six hundred members of his Lawrence local were working steadily on large-scale commercial jobs, including a four-and-a-half-year-long Western Electric project.

After years of cuts and freezes, wages started to climb. During the war years, carpenters in the cities of Boston, Springfield, and Quincy earned $1.50 an hour, only 12½ cents more than they made in 1929. In some of the outlying areas, the fifteen years between 1929 and 1944 had brought no gains. Carpenters in Hudson and Marlboro, for example, never received more than $1 an hour. By 1950, however, Boston carpenters were paid $2.37½, and the lowest rate in the state had jumped to $1.70.

The housing and commercial boom was accompanied by continued military construction. When the guns stopped firing in 1945, federal expenditures slowed briefly. But the new role of the United States as a world power and the onset of the Cold War moved Congress to beef up the military budget. During the 1950s, funds allotted for the construction of new military facilities escalated by 969 percent.[4] Wilbur Hoxie, safety engineer for the U.S. Army, remembers that Boston's building trades unions sent two thousand men over a seven-year period beginning in 1947 all the way to Limestone, Maine, to build the world's largest reinforced concrete arched hangar. A thousand carpenters built housing at Westover Air Force Base between 1948 and 1950. "I remember when we used to line up for payday," smiles Bob Jubenville. "The line would go all around between the houses; it was half a mile long." For many, the military work was the introduction to a measure of security in construction. Leo Coulombe worked as a foreman at Camp Edwards. From that time on, "I didn't have to ask for a job," he says. "They used to come to get me."

Jubenville and Mitchel Mroz, both of Greenfield, were among the state's many carpenters who entered the service and worked in carpentry shops in England and France during the war. Jubenville had joined the union in 1938; Mroz signed up when he returned, taking advantage of the union's offer of free admission to veterans. The homecoming of the nation's servicemen replenished the ranks of the locals, but the existing pool of skilled union carpenters could not keep up with the feverish demand for new construction.

Good times had arrived. Seasonal weather variations dictated that carpenters still averaged the traditional forty weeks of work a year in 1950, but wages were rising and jobs were plentiful. The problematic shortage of craftsmen was a boon to union members; their training and experience virtually guaranteed stable employment. Union strength was at an all-time high. Big projects were invariably built by union trades workers and even residential work was largely organized. A 1950 National Association of Home Builders survey showed that 65 percent of the employees of NAHB firms were union members. In Massachusetts, the union presence was even stronger. Arthur Anctil of Taunton, Joseph Emanuello of Weymouth, Enock Peterson of Framingham, and Thomas Phalen of Fitchburg point out that their areas had been solidly union well before the war. Boston and some other larger cities had long been union building towns. There are no adequate statistics, but most observers agree with Croteau of Lawrence that "anyone who swung a hammer belonged to a union."

But Croteau and many others believe that the very successes of the unions during the postwar boom laid the groundwork for their decline. Steady work made the members "fat and lazy," he says. "They forgot where they came from." Extra change in the pocket eased the struggle for security. Carpenters, once confined to urban working-class neighborhoods, now drove cars from the construction site to homes in the suburbs, far from the union halls.

Union members opted for commercial work. Housebuilding was hard, repetitive work. Residential contractors developed informal quotas, expecting a certain number of walls and roofs framed every day. It has always been that way. John MacKinnon worked on dozens of triple deckers in Dorchester during the 1920s for Gillis and McGillvery.

You put three men on one side of the house and three men on the other side and shingle that house. You shingle one side in a day from the bottom to the roof and if you were working for Gillis and McGillvery and it happened to be Gillis that day, he would say to you, "If you worked a little harder, you could put on the trim."

Industrial and commercial jobs were not necessarily easier, but their massive scale and individual designs made production quotas harder to pin down and enforce. Jobs lasted longer and, with hundreds of construction workers on some projects, supervision was often a little looser. "We found it much easier to work on these [jobs]," says Joseph Lia, General Executive Board member of the International. "Sometimes we could hide, disappear a little bit, nobody would know the difference. On housework, they could almost count the studs that went up at the end of the day."[5]

Housebuilding is construction's most volatile wing. Highly responsive to subtle shifts in demand as well as fiscal and monetary policies, the industry's erratic peaks and valleys have long been used to forecast national business cycles. Smaller residential builders frequently function on the margins, alternating between dreams of prosperity and fears of extinction. Employment in the residential sector, even in the halycon days of the fifties and sixties, has always been very risky. Individual jobs last weeks, not months. The feast-or-famine nature of the business makes long-term planning impossible. Most union carpenters were only too glad to leave that world behind.

In their rush to commercial projects, union

carpenters ignored developments in house building. A fully employed membership viewed residential jobs as second-class work. But the insistent call for more housing created a vacuum that needed to be filled. A new group of builders emerged. Some were former union men hoping to strike it rich as subcontractors for tract developers. As Don Danielson, former staff member of the International, put it, "A carpenter would say, 'I'll frame 'em up for so much a piece, I'll take eight or ten houses.' Then anything goes: one guy and four cousins, anybody can throw 'em up and sell it."[6]

There was also a generation of newcomers with no union tradition. Eric Nicmanis, now a nonunion framing contractor on the South Shore, arrived in the United States shortly after the war and worked on farms in West Virginia and New Hampshire. "Some friends in Boston came to visit us and said they had a lot of building going on and they need carpenters and they were making better money than we did on the farm." Nicmanis left New Hampshire, moved south, and began his framing career, putting up ten to fifteen house shells a year.

Nicmanis went to work for Campanelli, then the biggest developer in eastern Massachusetts. Many carpenters point to Campanelli's rise as the first break in union control of homebuilding. Arthur Anctil remembers when Campanelli bought huge chunks of land in Raynham, subdivided, and built the first large developments. According to Chester Sewell, "Campanelli had no use for the union. He came up here [Framingham area] and built all these gangs of houses."

Like most of the postwar developers, Campanelli revived the lumping system, paying individuals or small crews for particular tasks, such as framing or roofing. "He hired kids," says Sewell, "giving them so much to just board and roof with plywood." Campanelli managed to attract some union carpenters too. Though he paid no overtime, he encouraged workers to keep at it for as many hours a week

as they wanted to work. "It bothered me an awful lot that he was getting all of this work so that all of a sudden some fellows that weren't strong union members went with Campanelli because he gave them steady work. He was the one who really weakened the union here," concludes Sewell.

The nonunion homebuilders took root in the rural areas where the unions were weakest. Once established, they bid on jobs closer to urban centers. Ed Gallagher, of Local 275 in Newton, says Campanelli and other developers took over the residential work in his area by 1958. By the end of the decade, homebuilding in Massachusetts had reverted from an entirely union operation to a predominantly nonunion industry.

For the first time, commercial and residential wage rates for carpenters drifted apart. Large general contractors bidding on multimillion-dollar projects worried less about wage hikes than did small builders operating close to the vest. The big builders focused on completion dates and an adequate supply of skilled labor. With labor costs diminishing as a percentage of total construction costs, commercial contractors willingly accepted union demands for higher wages in exchange for qualified workers. By 1959, union carpenters in Framingham, Worcester, and the Boston area earned $3.40 per hour. Others across the state made slightly less, but still considerably more than unorganized carpenters. During the fifties, on the other hand, nonunion homebuilders either paid lump sums or $2 to $2.25 on an hourly basis.

In Boston, the nonunion homebuilder was a non-issue. The bulk of the city's jobs were in the commercial sector. Inroads into residential work mattered little. In other areas where it was more important, the unions expressed only slight concern. "We had plenty of work," says Joseph Emanuello, "so we thought we didn't have to picket." Most locals shared this attitude. "Everybody that I talked to said there's nothing that we can do," reports Chester Sewell. In isolated instances, unions tried to pressure nonunion contractors. The Brockton local picketed Campanelli, for example, but the glut of nonresidential work discouraged sustained campaigns.

In 1955, locals in Cape Cod, Cape Ann, and the Pioneer Valley instituted a dual wage system, one rate for residential and another for nonresidential work. In Greenfield, union contractors were required to pay $2.50 per hour for commercial work and 10 cents less for house building. In the next few years, the dual rate spread to other areas, including Lynn, Taunton, Springfield, and a number of towns in western Massachusetts. In some cases, the differential was minor, in others significant. In Franklin, the standard rate in 1968 was $5.50 per hour; carpenters who worked on jobs under $25,000, however, got only $3.55 per hour. By the end of the 1960s, the system faded away, ineffective at slowing the disappearance of unionized residential and small commercial work.

Dual rates had supporters and critics. Emanuello thinks the system came too late to do any good. Croteau agrees, citing an overwhelmingly apathetic response when he first proposed it in his local in 1950. By the time dual rates were instituted, both men believe the momentum had gone too far. The Greenfield experience is instructive. Nonunion builders entered the Pioneer Valley a few years after establishing beachheads in the eastern part of the state. The western Massachusetts locals had some time to prepare. Angelo Bruno claims, "We were pretty much on top of it and the split scale worked up to a point." Greenfield is located six miles from Vermont "in the fringe areas between unionism and nonunionism," says Bruno. "The work we had started to lose wasn't from the people in our area, it was the contractors coming down from up above. We kept some of the big ones in line by signing letters of intent and forcing them to abide by our conditions. Cushman was always nonun-

Angelo Bruno
By Nordel Gagnon

ion in Brattleboro, but when he came down to Massachusetts, we went to work on him and signed him up."

Employer abuse of the dual rates undercut support within the union. Contractors shifted carpenters from low-scale to high-scale jobs during the course of a single day, but paid the lower rate for all eight hours. "This kind of thing kept coming back to the hall," says Mitchel Mroz, "and people said, 'if that's the way the contractors are going to do it, we'll get rid of it.'" Abuse was not the only factor. The system was killed by the same dynamics operating in the rest of the state—apathy predicated on good times in the nonresidential sector.

Bruno stresses that residential work in his local "didn't really go downhill until all this work started coming through—Interstate 91 in the late fifties, the Rowe atomic plant in the early sixties. That's when the small stuff went out. Everyone went on the big jobs. We saw it happening, but no one cared." Mroz underscores the point. "It was the god almighty dollar. When all those big jobs were going, people could work all the hours they wanted. They even told the guys they could almost fall asleep on the job and still get paid for it. There were twelve-hour shifts and you could work twenty-four if you wanted. They offered that so they could get the help."

Support for the split scale gradually vanished. Union members felt it was abused and irrelevant; nor did the unionized employers care. The contractors on the Pioneer Valley contract negotiating committees were all big builders. None of them built houses. Union officers and rank-and-file carpenters recognized that homebuilding was falling out of their grasp, but the loss seemed incidental when stacked up against the surplus of better-paying, more secure nonresidential jobs. Occasionally, a voice in the wilderness cried out. At the 1965 state convention of carpenters, General Executive Board member Charles Johnson complained: "This organization was founded on home construction, and we are letting it go by the board."[7] But these periodic warnings were rarely followed by organizational action. Few union members felt a sense of urgency as long as union-controlled commercial work remained plentiful.

Roof framing, c. 1926. MHS–UBJCA

Rooftops and Train Tracks:
From the Twenties
to the Fifties

(*Above*) Although available from the 1920s, the electric circular saw did not take over on-site cutting functions until after World War II. GMMA

Wood scaffolding supports mason working on exterior of an MIT building, 1931. MIT

Depression-era federally funded construction on Huntington Avenue subway line in Boston, 1938. NA

Installing temporary tracks on the Huntington Avenue line, 1938. NA

Procter & Gamble building in Quincy under construction, 1939. NBM

MIT's Kresge Auditorium under construction, 1953. MIT

Concrete bucket, suspended from crane, about to release load into forms, Kresge Auditorium at MIT. MIT

(*Below*) Pouring the roof. At left, a laborer vibrates the concrete to remove air pockets. Carpenter, bottom right, watches forms to make sure they hold up under pressure from flowing concrete. MIT

Running the circular saw on
auditorium roof. MIT

Nailing the rafters. Photo A. M. Wettach/Black Star. GMMA

Giant beams cap wood-framed building. UBCJA

14

New Tools, New Materials, New Methods

"When I joined the union in 1943," Robert Thomas says, "the old-timers then were what I consider to be real true craftsmen. They were more or less used to working with their hands and their tools. It was a necessity. At that time they didn't have all the sophisticated power machinery that they have today."

Before World War II, hand labor dominated on-site construction. Door, window, and trim manufacture had long since passed to the factories, but carpenters still installed all the rough lumber and millwork with hand tools. Power equipment was available. In fact, from the time that Rudolph Diesel won a patent for the pressure-ignition internal combustion engine in 1893, gasoline-driven equipment had slowly made its way onto the job site. Hoisting rigs, air compressors, and primitive earth-moving machines took over arduous manual tasks and expanded the visions of designers and engineers. In its early stages of development, however, this equipment was bulky and inflexible and the prohibitive expense restricted it to larger building projects.

Bob Jubenville's first job was on the Sunderland Bridge across the Connecticut River in 1936. Building bridges is a major endeavor; the jobs have long life-spans and involve large numbers of workers. As a result, road and bridge contractors benefit from investment in labor- and time-saving capital equipment. The Sunderland contractor chose to experiment with state-of-the-art machinery, purchasing a gas-driven table saw for the carpenters. Jubenville reports that the boss's toy was a mixed blessing for the men who handled it. "They were dangerous. They had to be cranked up and the handles would come off sometimes. A guy broke his wrist because they'd kick back."

Mitchel Mroz also started his career in heavy construction, working with concrete forms. "On the big jobs, they had standard mixers for concrete. A truck would come with two batches of sand and stone. Then you'd throw in your cement bags—six of them. You put in the water, mix it up, and you'd have a [cubic] yard of concrete. You'd start at seven in the morning and sometimes it was ten or eleven at night before you'd get home on a 100-yard pour. The old-timers used to mix it all by shovel and hand."

The old days were not necessarily better; there was nothing glorious about concrete form work in the first third of the twentieth century.·

145

"Form work was all bull work. There weren't any cranes to pick things up. You know what they say, you had to have a strong back and a weak mind." A 1930 survey of the building industry by William Haber supported Mroz's remarks. Professor Haber archly noted that "strength and moderate skill in the use of hammer and rule are all that are required [on form work]."[1] In the late 1930s, trucks began carrying premixed concrete to the job, ending the fourteen-hour days of hand mixing. The trucks were small, handling 2, 3, 4, or 5 cubic yards of concrete at most. Since then, concrete manufacturers have learned to produce and transport larger batches. Today, dozens of 15-yard trucks regularly line up ready to deliver their load for poured-in-place highrise or highway projects.

Jubenville's prewar encounter with a power saw was unusual. Most carpenters, from rough-form builder to finish man, relied exclusively on hand tools. The widespread introduction of portable electric power tools after the war transformed the carpenter's day at work. The reciprocating saw, the drill, the power-actuated fastening gun, and above all, the circular saw (conventionally known by the brand name Skilsaw) forever changed how the carpenter cut, shaped, and fastened materials.

Many carpenters viewed the flurry of new tools with alarm. Still reeling from the calamitous Depression, few workers welcomed labor-saving innovations. Leo Coulombe admits that the Skilsaw made cutting easier, but he still worries, over thirty years later, that "it took work away from a lot of people." The older men, in particular, resented the new saw and opposed its use. Joseph Emanuello points out that some carpenters went to great lengths to express their hostility. "One guy took the Skilsaw on the roof of the building and threw it off. He said, 'The saw's too fast, I'm going to cut by hand.'"

Others accepted more readily. Thomas praises the role of power tools in the trade. "They speed up the job and take a lot of hard labor out of it." Paul Weiner agrees. "It's very positive. Instead of breaking your back cutting fifty boards in one day, you can do the work in a few minutes." Ernest Landry claims it was an adjustment, but he did get used to power tools and even learned to like them. Ultimately, he believes, the electric planes, drills, and saws "do a better job" than their manual counterparts.

In retrospect, it is hard to believe that the circular saw and electric drill, now such integral parts of the construction process, created such a controversy. Industrial workers had long since been forced to accept technological marvels that dwarfed the invention of the circular saw. At the same time as Joseph Emanuello's friend was throwing his saw off the roof, for example, computer-controlled machines were driving oil refinery workers to unemployment lines.[2] But the pace of technological change is a relative, not absolute, phenomenon. People respond to changes in their immediate environment, and carpenters had experienced no challenge on this scale to their work methods since the inception of factory-produced millwork in the late nineteenth century. Two, perhaps three generations of carpenters had learned their trade and passed it on with little change from one to the next. New and unfamiliar patterns cause fear. Leo Coulombe, Oscar Pratt, and many others of today's retirees remember quite clearly that the older carpenters of the fifties were frightened by the Skilsaw. The early models had only marginal safety features and often proved dangerous in the hands of a first-time user. Chester Sewell saw a workmate almost cut his leg off, and a sufficient number of similar horror stories circulated among the older carpenters to reinforce the built-in reluctance to disrupt their culture of work.

"We couldn't refuse to use them," Pratt points out, "even though we could see they were going to knock the heck out of the trade." But, he notes, they were expensive and that created another problem. Skilled workers took pride in their tool collections. "Every available carpentry tool that you could get, you'd have," asserts Joseph Emanuello. The tool chest traveled to the job site with the carpenter. But with the prohibitive costs of early power tools, jobs would go only to the men who could afford the price if a contractor expected carpenters to supply them. The unions quickly recognized the potential dangers. As early as 1932, the Springfield District Council proposed fining union members $25 if they furnished electric hand-driven machinery to a contractor.[3]

By the time power tools inundated the field, the unions had successfully spelled out employer responsibility. Union by-laws contained provisions requiring employers to supply whatever power tools, levels over 30 inches, long ladders, sawhorses, and extension cords were used on the site. The occasional infractions only underscored the stakes involved. In 1964, a Chicopee contractor told a union business agent that the work of Donat Charpentier, member of Local 685, was unsatisfactory and tried to prove his point by referring to the superior speed of another union worker, Philador Lemay. Upon investigation, it became clear that Lemay's superiority rested on his willingness to use his own truck, 4-foot level, and electric saber saw for company purposes. Lemay was found guilty of violating union rules and fined $25.[4]

The Skilsaw symbolized a loss of control and a new order, but it brought external changes as well. The traditional image of the nattily dressed carpenter whose white shirt and bow tie were covered by a pair of white overalls gradually disappeared. The overalls had served a dual purpose: they protected the layer of clothes underneath and held an astonishing array of hand tools in dozens of pockets and openings. The power tools made some of the hand tools obsolete or redundant, and the modern carpenter found he could carry all the tools and nails he needed in a simple apron or pouch.

The power tool invasion was accompanied by a host of new materials. Escalating wood costs meshed with shifting architectural tastes to simplify design and installation techniques. Oil-based products, like rubber baseboards and plastic casings, superseded complex wood moldings that had been fabricated on the job from multiple pieces of millwork. Plywood replaced tongue-and-groove boards in concrete forms, house sheathing, and sub-flooring. Asbestos and aluminum siding challenged the centuries-old exteriors of wood shingles and clapboard.

Heavily wood-framed multistory buildings, already virtual dinosaurs, completely disappeared. Concrete, the preferred medium for commercial construction, was not a wood product but its erection still required dozens of carpenters on a big job. The development of precast concrete—giant slabs of factory-made concrete shipped to the site—cut into the work load. John Greenland remembers the first time he saw a wall of precast concrete at the Howard Johnson's in Boston's Kenmore Square. "The framer came in, took a big goddamn side off the truck and put it up and in. It was amazing." Like precast, the use of steel as a structural building block reduced the need for carpenters on the exterior frame.

Inside the buildings, walls were covered by sheetrock, a gypsum-based product produced in large rectangular panels. Sheetrock quickly became the industry standard, first in residential construction, then throughout the industry. A 1956 study by William Haber and Harold Levinson estimated that 95 percent of all new houses built in the 1930s contained a plaster finish on the interior walls. Plastering was a time-consuming process. Angelo DeCarlo, who worked on dozens of houses in the Jamaica Plain section of Boston in the twenties and thirties, suggests that finishing plaster walls regularly took two weeks for each unit. During the first week, carpenters nailed small strips of wood lath across every stud to hold the wet plaster; the following week, the plasterers took over, applying first the base and then the finish coat of plaster. On the other hand, sheetrock (or drywall, as it is also known) was installed in a matter of days, requiring only a few thin coats of joint compound or spackle before painting. By 1952, Haber and Levinson claimed that 50 to 60 percent of all new homes used drywall.[5] Today, the plaster technique is reserved for restoration or specialty purposes.

Drywall is heavy work. Individual sheets weigh from 50 to 100 pounds depending on size and thickness. At a time when established carpenters had attained a measure of security with somewhat less demanding work, few of them jumped at the chance to learn the new branch of the craft. Contractors relished drywall because the standard sheet sizes (4' × 8', 4' × 10', 4' × 12') and the repetitive character of the work made the volume easy to measure. A trade based on speed, production quotas, and constant wrestling with heavy materials attracted few carpenters. The very makeup of the product—two layers of thin paper sandwiching crushed gypsum—only confirmed the distaste of men used to handling wood.

Nonetheless, sheetrock graced the walls of most new houses in the fifties and someone had to put it up. Harold Humphrey, who has worked with drywall ever since it became popular in the Northeast, says the vacuum was filled by a new influx of French Canadians. Initially, they came for a few months at a time, saved their earnings, and returned to Canada. Brothers, cousins, or groups of friends traveled together and worked as a team. After the framing carpenters had completed their work, the sheetrockers whipped through a house in a cloud of gypsum dust, paid either by lump sum or per piece of sheetrock. Eventually many of them settled in New Hampshire, Vermont, and sections of Massachusetts. Today, much of the drywall work in Massachusetts is installed by first-generation French-Canadian contractors and workers.

Humphrey says "the older American carpen-ters never got the knack of drywall. Since they thought it was too heavy, they were licked right there." Ernest Landry admits that he held that view. He refused a job when told he was expected to "hang" 60 sheets a day. As a finish carpenter, he told the foreman, "That's not for me." Leo Bernique, who has spent much of his life on highly skilled finish work, contrasts two jobs he held in 1983. At Suffolk University, the work was careful and demanding and he loved it. On a job at Boston University, he worked in a sheetrock crew. "It was tedious," he reports. "There was no challenge. You couldn't breathe. You were just making a day's pay."

Humphrey thinks many carpenters have lived to regret their disdain for drywall. Metal studs have displaced wooden studs in the partition walls of commercial buildings and that framing system has been incorporated into the work of the sheetrocker. Over the years, drywall contractors have emerged as significant sources of employment for nonresidential as well as residential carpenters, crowding out other employers who work strictly with wood. Barney Walsh, business agent of Boston's Local 67, estimates that drywall now accounts for a third of the total work in his area. A generation of drywallers has now perfected the installation methods, moving quickly with as little wasted motion or material as possible. All-round carpenters or those who are skilled in other parts of the craft have little chance of competing with a drywall specialist. Bernique acknowledges, "The French Canadians do it year round. You can't beat them."

Drywall may have been the biggest, but it was not the only sub-specialty of the trade to appear in the postwar years. "When I learned the trade," notes Oscar Pratt, "I learned it from the footings to the last piece of finish that went on a building. Since the 1940s, the new materials started to break the trade down." Bob Thomas thinks the push toward specialization intensified in the sixties. "You'd begin to see one man set door bucks, another put in

Harold Humphrey
By Nordel Gagnon

Bob Thomas
By Nordel Gagnon

hurt the union. As the craft becomes an amalgam of sub-trades, the common thread of the carpenter's identity is weakened. Many of the specialty contractors, especially in drywall, arrange to pay on a piecework basis. They skirt union regulations against piecework by issuing one check based on the union hourly rate plus a supplementary "bonus" check that covers the difference. Some of the fastest sheetrockers are able to take home well over $1,000 per week. With those earnings, even workers who carry a union book are unlikely to complain. And, as Croteau observes, the union cannot successfully police violations without cooperation from the members.

Ed Gallagher agrees that the splintering of the trade means carpenters "owe their allegiance to fewer people." Many of the specialized skills, such as ceilings, floor laying, and drywall, are performed indoors. In a healthy economic climate, these workers can be employed fifty-two weeks out of the year. They rely less on the union to cushion seasonal layoffs and more on individual employers for a stable relationship at a piece rate. Specialization has, in effect, sharpened a chronic division in the industry between "company men" (those who identify their fortunes with a particular employer) and the bulk of working carpenters whose overlapping loyalties may include the union, various contractors, and an amorphous craft identity.

In an environment riddled with insecure employment opportunities, workers inevitably struggle to find stability. A common solution is to seek less lucrative but more reliable carpentry maintenance work in hospitals, schools, factories, or other large institutions. Others leave the construction industry altogether, settling into alternate occupations. For those who stay in the trades, hopes of stability are often tied to the search for a permanent employer. Ernest Landry worked for five years with the Salem-based Pitman & Brown. He recommends "hooking up with a contractor

cabinets, another put up acoustical ceilings, or lay floors." The longer carpenters trained on a particular task, the more efficient they became. Contractors grew reluctant to lose speed with nonspecialists. "You do something long enough," comments Thomas, "and you got to be clever at it."

Richard Croteau thinks specialization has

and staying with him. Every job he takes, you're one of the men and you know what to do."

In many cases, the goal is to find a "good" employer, that is, one who builds and retains a regular crew rather than constantly hiring and firing. Arthur Anctil built houses, schools, and churches for Witherall from 1923 to 1940. The business, he says, operated like "a family." Anctil was elected to a number of union offices over several decades, and despite his strong identification with the union, he still refers to his former employer as "we." Leo Coulombe had a similar experience in his fourteen years with Harry Thorley, a local New Bedford contractor. "He was one of the best, a gentleman. If you were sick, you never lost a day's pay."

A long-term relationship with a contractor is often viewed as recognition of competence and trade skill, especially if it leads to periodic stints as foreman or even superintendent. Yet the term "company man" has a decidedly pejorative ring. Though some craft knowledge is clearly indispensable in order to move from job to job, it is a common observation in the industry that hiring is done on the basis of "who you know, not what you know." Extended ties to a contractor are often judged as excessive loyalty, insufficient independence, and an uncritical acceptance of company practices. Nor are these views necessarily the sour grapes of those who never "made it." Joseph Emanuello worked for O'Connell for a number of years, but he is quick to point out that "I always considered myself a union man, never a company man." Tom Phalen spent fourteen years with W. J. Hanley until the Fitchburg contractor ran out of work during World War II. After two years at Fort Devens, he chose not to return to Hanley so he would not "feel obligated or part of the family, because local contractors sometimes thought they owned you."

Even carpenters like Landry who strongly favor finding a permanent employer caution against harboring illusions. "If you didn't do

exactly what a lot of contractors said," he argues, "you'd get through at night. They used to hire twenty men in the morning and fire twenty-one at night." Joseph Petitpas supports Landry's contention. "Years ago, Coleman Brothers and them used to hire fifty guys in the morning, lay forty-nine off at night, hire fifty more the next morning, and lay forty-nine off at night until they got the crew they wanted. When the guy wasn't producing, down the road he'd go. That's why the union started saying, 'Look, we're giving you forty men and we're not sending you another forty.'"

A 1960 Harvard University study of the Boston building trades demonstrates the hazards of "company man" aspirations. Though one-third of those employed had company connections, they were not necessarily regular employees. In fact, the survey of thirty-five general contractors (who hired laborers, bricklayers, and supervisory personnel as well as carpenters) showed that only three bothered to offer steady employment to as much as one-quarter of their annual labor force. And even those rare demonstrations of loyalty came from very small firms with total employment figures of eight, fifteen, and thirty. The larger contractors (the fifteen who employed a hundred or more workers a year) who dominated the industry were less reliable. Less than 6 percent of all their employees could be considered "company carpenters," i.e., employed on a year-round basis.[6]

In the unionized sector of the industry, an alternative strategy was to be part of the union "in crowd." Since many requests for manpower came through the union hall, active union members and friends of the business agent counted on relatively steady employment. Perhaps they worked for many different employers, but new job placements quickly followed layoffs. This quest for security also had its critics, convinced that jobs were won through friendship and political ties rather than merit or skill. Angelo DeCarlo worked for

Abel Ecklov from 1917 to 1941 and had little use for the referral system. "They always had their favorites," he comments. Al Valli agrees that business agents practiced favoritism, but is not as critical. "It's only natural to take care of your friends," he says. When he was unemployed, he would call his agent. "If he had a place to put me, I'd go and if he didn't, I'd go find my own."

The unions remain the most reliable sources of employment. A union card has been a requirement for carpenters on large-scale jobs in Boston for almost a century. That does not mean, however, that union hiring halls have supplied all the labor power. Barney Walsh says that his local refers 25 to 30 percent of the carpenters on union jobs in his jurisdiction today. According to Abraham Belitsky, author of the Harvard study, similar conditions prevailed twenty-five years ago. His survey of building employers indicated that one-third of their labor force came from union referrals, one-third from previous employment with the company or personal references from a foreman, and one-third were hired off the street at the job site.[7] Oscar Pratt, former business agent of Brockton Local 624, claims that an increasing number of jobs have gone through the union hiring halls as the proportion of large contractors has grown. Bob Marshall, business agent of Boston Local 33, sometimes prefers to deal with national contractors exactly for that reason. "They're in for a short period and get all their employees right from the hall." Pratt says the locals have tried to discourage carpenters from finding their own work. The union allows individual solicitation, but he argues that the presence of job seekers on the site encourages employers to lay off their existing set of employees and institute the "revolving door" hire-and-fire policy.

Carpenters could always fall back on finding their own work. "I always got my own jobs," claims Leo Coulombe. It was a matter of pride for him not to go through the union. The op-

tions varied from town to town and year to year. In the twenties, Enock Peterson recalls, "we had a business agent but there was no office. So you'd go out and get your own job." Carpenters moved to where the jobs were, in or out of state. Travelers shared rented rooms and food expenses, hoping to get home to their families on the weekend. Croteau describes the classic method of finding work. "You'd get three or four guys together in a car, drive around watching for trucks with loads of lumber. Then you'd follow it to the job and ask for work."

Construction is a notoriously unsentimental business. Despite the occasional accolade to the "good" employer, most carpenters today observe that "you're just a number." The loyalty that Anctil and Coulombe cite has waned with the succession of regional and national contractors over local employers. Local contractors shared communities with their employees. Criticisms affected their reputations and their ability to land contracts and find skilled workers. The Belitsky dissertation, published in the transitional year of 1960, concluded that local firms frequently "carried" men through slack periods. In contrast, national contractors hired and fired strictly on the basis of their needs.[8]

For several years, Bob Thomas worked for Grant, a small contracting company. "I never lost a minute of work. If things were slow, we'd sit in the office. It's completely different with big outfits. You're a name on the payroll. If you're doing the job, fine. If you don't, they'll get somebody who will. That's the way it is, from the top echelon all the way down the line." Tom Harrington remembers when one of his father's employers would visit the family on Sunday for home brew. "Contractors used to be close to the men. Now it's big offices and big oak tables. The only time a man working outside knows who his big boss is is if he sees it on a sign or in a newspaper." Ed Gallagher, business agent of Newton Local 275, says this trend has had an effect on relations with the union. "There used to be more personal contact. You'd know the owners on a first-name basis. Now the agent meets with the project manager to handle problems."

The hunt for complete security in construction employment is futile. There is no informal tradition of job tenure, nor has any collective bargaining agreement ever included provisions for seniority. Construction is one of the few organized industries in which unions have never challenged the employer's unilateral right to lay off without notice. Contractors reg- ularly dismiss long-term employees for reasons as arbitrary and diverse as advancing years, declining health, injury, company reorganization, personality conflicts, or most commonly, lack of work. The various individual strategies to ward off the anguish of insecurity—union man, company man, or a protective demeanor of independence—are fragile reeds at best. With no institutionalized protections, carpenters have been forced to follow their own paths, knowing what slender threads their livelihood hangs on.

15

The Prudential Boom
and Beyond

The 1950s had been good to union carpenters. From 1950 to 1960, wage hikes across the state ranged from 33 percent to 42 percent. Other building trades workers fared equally well, and in fact, successful contract negotiations in one trade often boosted expectations in another. In 1956, Boston carpenters went on strike after rejecting an Associated General Contractor (AGC) and Building Trades Employers Association (BTEA) offer of a 30-cent raise over two years. During the walkout, carpenters learned that Boston's bricklayers had signed a one-year contract calling for a 25-cent raise. Refusing to be outdone, carpenters' officials promptly adjusted their demands upward from a 35-cent to a 45-cent package over two years. A union spokesman declared: "We feel the difference between the carpenters and the bricklayers is large enough right now. We don't intend to stand by and let the gap be widened any more."[1] Rejecting offers of mediation from state officials, the union continued the strike until the employers relented, agreeing to a 40-cent raise.

Boston's locals were not quite as successful in negotiations during the 1958 recession. Despite employer warnings that they would use rising joblessness (almost three thousand Mas-

sachusetts construction workers had exhausted their unemployment benefits) to hold the line on wages, union members voted 4 to 1 to authorize a strike for a 50-cent package over two years. On May 12, sixty-five hundred carpenters walked off $100 million worth of construction projects. After twenty-one days and the intervention of International officers and federal and state mediators, the union settled for a 40-cent increase over three years.[2]

Before the war, bargaining a contract had been fairly uncomplicated. "Back then, all it amounted to was wages," says Carl Bathelt. "You didn't even need an agreement." In fact, there were no written contracts in many of the state's smaller towns. In the 1942 State Council's booklet of wage scales, only half of the thirty-eight local unions and district councils indicated that they had written agreements. Seven areas had oral understandings and the remaining twelve set their wage rates without an agreement of any kind. Contractors haggled with union representatives, but a handshake at the end of the sessions was enough to govern the coming year's labor relations. As late as 1956, locals in Attleboro, Clinton, Franklin, Lawrence, Northampton, and Westboro operated without bothering to sign a piece of paper. Even when they were recorded, the agreements were simple. "When I took over as BA in 1954," says Thomas Phalen of Fitchburg Local 48, "the contracts were one page long, and everything was on one side of the paper."

Once the unions began to expand their negotiating agendas, the system of informal agreements was obsolete. "Before then, you got your wages, period," comments John Greenland. "Then we started to get more social benefits." Following the lead of the needle trades and miners unions, building trades officials called on employers to establish funds for illness and retirement benefits. In 1948, the Boston locals of sheet metal workers, asbestos workers, roofers, and electrical workers established health and welfare funds. In Springfield, the operating-engineers union negotiated a similar fund the following year. Ernest Johnson, secretary-treasurer of the Building and Construction Trades Council of Metropolitan Boston, labeled these funds "one of the greatest union gains in many years."[3]

In 1950, negotiators for the carpenters proposed a health and welfare fund and a pension fund to the AGC and BTEA. The contractors firmly rejected the ideas, arguing that construction workers moved too freely from one employer to another and in and out of the industry to make such agreements feasible. Phil Conte, who has participated in decades of bargaining sessions as an employer representative, remembers the initial reaction as "absolute terror and anger." As chairman of a committee on the proposal, Conte disagreed with his fellow employers, suggesting it was "not a wild dream of the unions." Most of the contractors, he remembers, took a narrower view, claiming that the funds were "not necessary to put up a building." The proposal was set aside for further discussion.

In 1952, the Springfield District Council of Carpenters and the Northern Massachusetts District Council (representing the Fitchburg-Leominster area) signed agreements with their employers for 7½-cent hourly contributions to a health and welfare fund to be administered by a joint union-management board of trustees. In their negotiations, the Northern Massachusetts Council also won an employer-financed vacation fund. In 1955, the Boston District Council, Brockton, Fall River, Newton, Middlesex District Council, and the Norfolk County District Council joined the pioneers with their own health and welfare programs. Ten years later, every local in Massachusetts belonged to one of the state's twelve funds. The Boston fund opened with a $300 life insurance benefit, a $10 daily hospital room allowance, a $600 surgical fee, and an $80 maternity benefit. Since then benefits have risen with hospital costs and now include dental programs and the services of a vision and diagnostic center in Cambridge.

By 1962, eighteen of the state's twenty-eight agreements included a pension fund as well. Tom Harrington was on the negotiating committee of pile drivers Local 56 that year. Winning the pensions took "a hell of a big fight," he says. "The contractors called us socialists and communists." Oscar Pratt negotiated for the Brockton local. "There wasn't an employer who didn't fight anything of that kind," he reports. The Brockton contractors accepted the proposal only when they realized they could deduct their fund contributions as business expenses. New Bedford carpenters paid a steeper price, striking for two months before agreeing to forgo a wage raise in exchange for the funds.

Contracts also incorporated language that gave carpenters greater control over the pace and quality of their work. Nowhere has the tension between the contractor's goal of maximized production and the carpenter's desire for craft pride exploded more intensely than in the age-old conflict over speed and rest breaks. In the old days, Richard Croteau observes, many contractors would not hire carpenters who smoked pipes because lighting and relighting the tobacco took too long. Bob Weatherbee's memories are similar. "When I came in, we were like dogs. If you lit a cigarette, you were fired." A cup of coffee provoked the same response. "If you brought your own," says Joseph Emanuello, "you had to sneak it. You'd get fired if you were caught."

The demand for a coffee break turned into a struggle to preserve the workers' dignity. Tom Harrington describes the lengths to which carpenters went before coffee breaks were accepted as part of the daily routine. "You had to sneak it," he says. "A laborer would get it. You'd get behind something, drink your coffee down, scald your mouth, dig a hole, and bury your cup." Union negotiators found winning a coffee break to be as tough as any issue they had fought for. "We had one devil of a time to get it in the contract," Harrington continues. "The contractors would say, 'Nothing doing, nothing doing.' They wouldn't have it. When we finally won it, business agents had to go around *en masse* to make sure we actually got it."

The 1958 agreements included a five-minute morning coffee break for the first time. Contracts from this period include long-standing as well as recently bargained job practices. Travel pay, show-up pay, and double pay for overtime protected the wage standards; employer responsibility for sharpening tools protected the handsaw dulled by concrete form work; and drinking water and toilet facilities protected the comfort of the carpenter. Enforcement of these provisions, especially working conditions, has proved to be another matter, but the contracts of the fifties ushered in a new set of expectations for union carpenters.

━━━━━━━━

"Since the 1880's," writes Steven Miller, "Boston's manufacturing story had been one of uneven, relentless decline. . . . Heavy industry always had been located outside of the capital city, first in the satellite cities of Lowell and Lynn, and later in other parts of the country and the world. The port was strangled with the emergence of New York as the major east coast shipping center, and was dead by the turn of the century. Even the small industries that grew on the easily exploited Irish labor force after the Civil War died or moved out."[4]

The city retained enormous wealth and resources, particularly in financial and educational fields, but the region's Brahmin elite had blocked substantial capital investment in Boston for decades. The rise of the Irish political machine, symbolized by the legendary James Michael Curley, had provoked the undying hostility of the city's former political rulers. While construction in many American cities picked up in the late 1930s, "the negative attitude of the Yankee-dominated insurance industry was so fervent that no mortgages on buildings in Irish-dominated Boston were granted."[5]

Curley's political downfall in 1949 was engineered, in part, by a group of professionals and business executives who called themselves the New Boston Committee. Ten years later, attorney Charles Coolidge and banker Ralph Lowell initiated an even more aggressive busi-

Tom Harrington
By Nordel Gagnon

ness posture. Not content to defeat enemies, Coolidge and Lowell formed the "Vault," a publicity-shy organization of top business leaders, to reassert corporate influence over the city's political future. With Vault support, John Collins was elected mayor of Boston in 1959 and quickly moved to involve the business community in city affairs. Mayor Collins, along with Coolidge, recruited New Haven city planner Edward Logue to administer the newly established Boston Redevelopment Authority (BRA) and oversee the physical transformation of downtown and the renaissance of the city as a financial center.

The first landmark of the Collins-Logue administration was the Prudential Center, built on top of the former switching yard for the Boston and Albany Railroad in Back Bay. The design of the 750-foot tower introduced a new scale of construction activity to Boston. A total of 660,000 bags of cement were poured into the Prudential foundations and 60 million pounds of structural steel were fastened with 400,000 high-strength bolts.[6] Bob Thomas, along with most of his contemporaries, thinks the Pru marked a watershed in Boston's building history. "It made a vast difference in the trades," he insists. According to Thomas, there were at least three thousand union carpenters working in the downtown Boston area at the time. Local 33, with a membership averaging just over six hundred in the fifties, was overwhelmed by the boom. Cliff Bennett, business agent from 1958 to 1973, reports that he called unions across New England and ultimately welcomed hundreds of men from across the border in Newfoundland in order to supply the Prudential's ravenous appetite for carpenters.

From 1958 to 1959 (the year of the Prudential's ground breaking), the value of new office building leaped from $16 million to $68 million in Boston.[7] Construction workers streamed into the city in search of work. They found jobs in every corner of the downtown area as the boom continued into the 1960s. The

city's skyline moved ever higher. The Prudential was followed by the 60-acre Government Center on the sites of Scollay and Bowdoin squares, along with a barrage of banks, offices, and hospital buildings.

The unprecedented demand for labor encouraged the unions to call for large wage increases. Union officials held out threats of work stoppages willingly, knowing that contractors were reluctant to interrupt mounting business activity. Only a fifteen-hour marathon bargaining session averted a strike in 1963, as the AGC and BTEA agreed to a $1.05 package in exchange for five years of labor peace. In 1969, thousands of carpenters throughout eastern Massachusetts walked off $1 billion worth of construction projects. After forty-four days, negotiators agreed on a three-year contract that promised $8.80 in wages and fringe benefits by 1971. When the politicians had gathered for ground-breaking ceremonies at the Prudential site, Boston carpenters grossed $136 a week. In the 1969 settlement, after a decade of continuous building, the final raise brought their weekly paycheck (including benefits) to $352.

The redevelopment of Boston triggered building across the state. From the roadways of the Massachusetts Turnpike to the buildings of the University of Massachusetts campus, construction workers reshaped the physical appearance of the commonwealth. From 1961 to 1972, the construction labor force in Massachusetts grew by almost twenty-six thousand. In the first half of the decade, the rate of construction growth exceeded the national average and the pace only accelerated in the second half. From 1967 to 1972, building employers in the state raised their revenues 79 percent, up to $4½ billion.[8]

Carpenters look back fondly on the boom-time spirit of the 1960s. It was one of those rare periods when individual workers held the upper hand as employers competed for a limited pool of skilled labor. Supervisors often looked the other way at a slip-up, a mistake, a curse hurled at a foreman, or other actions once considered certain grounds for immediate discharge. Lay-off day, usually a source of dread and anxiety, turned into just another working day. A telephone call to the union hall or a former employer almost certainly produced another job assignment. By 1969, construction unemployment had dropped to 5 percent nationwide, the lowest point in twenty years.[9]

Quitting a job before its completion, previously unthinkable, became an option as men hunted overtime pay and better working conditions. "If a foreman looked at you cross-eyed," says one older carpenter, "you'd pick up your tools and go to another job down the street." For the first time, workers were able to choose jobs based on desirability, not just availability. Mitchel Mroz was one of hundreds of carpenters who worked at the giant Turners Falls Dam project in the western part of the state. "The Dam couldn't get enough help, so they went on a ten-hour day, five days a week. Then guys would hear about it and they'd leave their other job to make more money. Then another contractor would put more hours on, and guys were jumping around from job to job. If they could make more, they'd quit and move on."

The experience in Massachusetts reflected the national surge in the fortunes of construction workers. Wages and prices were rising in the inflationary Vietnam years, but none more quickly than in the building industry. By 1970, construction settlements were running at an average of 15 to 18 percent a year, while manufacturing agreements were jumping at no more than half that rate.[10] Contractors were thriving as well. Bankruptcies, always a chronic problem in the volatile industry, hit a fourteen-year low in 1969, and total construction earnings climbed each year to a record peak in 1973.[11]

There had always been voices of doom in the midst of the general contentment. Warnings of impending crisis stemming from a contracting labor force and expanding demand had been sounded for years. The low birthrate of the Depression years created a dip in the population of adults available for all industries in the fifties and sixties, and construction had been particularly hard hit. With increasing postwar employment opportunities, fewer new entries into the labor force were attracted to the demanding and insecure world of construction. Younger tradesmen were not replacing the retiring ones in sufficient numbers. As a result, there was a proportionately higher number of construction workers aged forty-five to sixty-five than in other industries.[12]

When a 1966 federal study forecast a need for 670,000 new construction workers in the upcoming decade, alarms rang in every corner of the industry. *Engineering News-Record,* the industry's trade journal, instituted a regular survey of labor shortages by trade and city. Large contractors, labor relations consultants, government policymakers, and major industrialists with large capital commitments in construction projects predicted spiraling building costs. Construction spokesmen inaugurated a campaign saddling the building trades unions and their restrictive admissions procedures with responsibility for the manpower crunch. They accused the unions of taking unfair advantage of labor shortages and argued that the resulting pressure on wages was producing a new unproductive atmosphere on the job site. A Detroit builder dusted off the old saw connecting productivity and low pay, claiming there was "an inverse ratio between the amount of wages paid and the will to work."[13]

The rhetoric quickly escalated. Editorial writers pointed the finger of blame at the unions not only for labor shortages, but for all outstanding problems in labor relations. In 1966, construction management consultant John Garvin set forth a program to "reform" collective bargaining arrangements and "eliminate the present imbalance of power" favorable to labor. Garvin proposed regional negotiations to strengthen the hand of scattered local contractors, the elimination of union rules prohibiting new technology, and the replacement of union hiring halls with employer-controlled data banks of workers.[14]

Journalists picked up the drumbeat. A *Fortune* article by Thomas O'Hanlon titled "The Unchecked Power of the Building Trades" attributed rising costs to the "murderous bargaining strength . . . [of] the most powerful oligopoly in the American economy" and its "stranglehold on construction." O'Hanlon's remedies paralleled Garvin's. He suggested that a traumatic confrontation with the unions would be necessary "if the industry is to become rationalized." The *Wall Street Journal* and other business publications also endorsed Garvin's ideas and called for quick action.[15]

In 1970, the Bureau of National Affairs published a book by M. R. Lefkoe, a business consultant on the payroll of several large construction firms. Lefkoe's work, *The Crisis in Construction: There Is an Answer*, recited the by then standard litany of complaints but went on to address the heart of employer concerns, the fear of loss of management control. Lefkoe recognized that construction workers had used promising economic conditions to win a measure of control and autonomy over daily life on the job. This, more than any other, was the issue that disturbed construction management. "Skilled craftsmen," he wrote, "knowing that they are so much in demand that they can always find work, have no qualms about not reporting for work whenever they please, or leaving one job for another if they don't like their working conditions or anything else about a job."[16]

According to Lefkoe, contractors had effectively relinquished authority to the unions over matters as diverse and important as the regulation of the labor supply, job assignments, output quotas, crew sizes, and the introduction of labor-saving machinery and prefabricated materials. "As long as contractors continue to accept the theory that the jobs they create *belong* to the men they employ to fill them (and

to the unions that represent those men)," he continued, there would be little hope for a return to the proper managerial role.

Contractors publicly paraded their "helpfulness" in an effort to pin responsibility for mounting building costs on the unions. After settling a sizable wage package, Frank White, executive vice-president of the Connecticut chapter of the AGC, described the bargaining process: "They demand, we give, and they take." Long strikes failed to moderate union demands as individual contractors inevitably broke with employer associations, rehiring workers on the basis of national or interim local agreements. Tradesmen often found work with contractors eager to get a leg up on their idled competitors. Workers therefore patiently bided their time until the recalcitrant builders were forced to give in. "We didn't negotiate," complained a Miami member of the AGC. "We pleaded. With an excess of construction activity and a serious shortage of manpower, too many builders didn't care what they were paying."[17]

The contractors' laments failed to shift responsibility. In fact, the opposite occurred. Critics outside the industry simply added another target to their broadsides. "Chaos in the Construction Industry," a 1969 report from the National Association of Manufacturers, assailed the twin problems of "the excessive, collective economic strength of the construction unions and the reciprocal lack of bargaining power of contractors and contractor associations."[18] In his book, Lefkoe had warned that "if companies from within the industry fail to take appropriate action," someone else would. As a spokesman for a group of large builders, Lefkoe's recommendations revolved around expanding the functions of contractors. Other forces in construction, however, had made it clear that they were skeptical of contractor ability to rein in the unions.[19]

Preliminary actions had already been taken. At a national conference on construction problems, Winton Blount, president of the United States Chamber of Commerce, described labor conditions as "chaotic and unbelievable, completely out of hand." He placed equal blame on the unions and on a decentralized management that simply passed increased labor costs on to customers.[20] Blount, the National Association of Manufacturers, and other industrialists questioned contractor interest in slowing union wage demands. While sympathetic to contractor concerns of managerial control, they first wanted brakes on their mounting construction bills.

In 1969, Roger Blough, one-time chairman of U. S. Steel, formed the Construction Users Anti-Inflation Roundtable to monitor and intervene in the building industry. The original policy committee of the Roundtable (renamed the Business Roundtable in 1972) included the chief executive officers of General Motors, General Electric, Standard Oil, Union Carbide, AT&T, and Kennecott Copper. Since then, the organization has come to include some two hundred of the nation's top chief executive officers whose companies spend an estimated $100 billion on capital projects. The formation of the Roundtable altered the terms of the discussion as a new set of players moved to wield their considerable influence. It was the most dramatic and organized entry of powerful construction clients into the workings of the industry since Boston's "gentlemen engaged in building" locked out the city's carpenters in 1825.

These modern-day gentlemen alternately expressed concern and contempt for their building employer brethren. Ready to find common cause against the unions, the Roundtable leaders were not, however, prepared to accept contractors as equals in wealth, power, stature, and above all, management sophistication. Initial Roundtable policy statements continually harped on construction's underdeveloped state, describing it as "differ[ing] in important respects from conventionally structured industry, where management manages." Their first report rejected the "peace at any price" approach, sternly lecturing weak-kneed contractors: "There is no substitute for firmness in collective bargaining." They chastised builders for using overtime to attract labor, arguing that regularly scheduled overtime magnified labor shortages, reduced labor productivity, and created excessive inflation. They attacked the union hiring hall system, concluding that any benefits were outweighed by "badly handicapping the ability of the contractor to manage his work force." Reforming the referral system was a hopeless task, the report stated. Like Garvin, Lefkoe, and others, the Roundtable advocated an employer-developed data bank of craftsmen. "Such a system would represent a long step in the direction of restoring to construction industry employers the control of the recruiting and hiring process that employers in other industries traditionally exercise."[21]

The Roundtable's authors called for the destruction of customary standard wage-skill structures and the development of new categories of multiskilled, semiskilled, and unskilled workers. Contractors, they charged, have blindly accepted archaic trade practices and union-defined craft jurisdictions, thus granting, in effect, "management-approved status." Without the restoration of genuine managerial authority in hiring, training, output, pay scales, discipline, and supervision, the report concluded, the union hiring hall would continue to "very substantially usurp the employer role normally reserved to management in other industries."[22]

The first Roundtable report, released in 1974, kicked off a $1.4-million study of the industry. Though this massive volume of paper recommendations carries no binding weight, it has been and is being used effectively as an organizing tool. In December of 1982, Roundtable chairman Charles Brown (formerly of Du Pont) convened a meeting of the fifty largest contractor associations to open the operational phase of the report. Brown told the assembled build-ers that contractor-owner cooperation was the key to thorough reform. If they were "only moderately successful" at implementing the suggested changes, he predicted the industry could save $10 billion annually.[23] The Round-table proposals have clearly emerged as the guidepost for governmental construction pol-icy and the industry's collective bargaining ses-sions.

The national building trades unions, some-what awed by the Roundtable's collective in-fluence, have treated the powerful organization gingerly. Joseph Maloney, secretary-treasurer of the AFL–CIO's Building and Construction Trades Department, described the unions' accommodating stance to a 1984 Roundtable conference. "I submit that we responded not at the speed of a glacier," Maloney said defen-sively, "but by making more changes in local labor agreements than were made in the pre-vious 80 years." Union leaders occasionally blast Roundtable antiunionism for the benefit of the rank and file. On a day-to-day basis, however, every national union has accepted the Roundtable framework. The much bally-hooed National Market Recovery Program, issued jointly by the AFL–CIO Building and Construction Trades Department (BCTD) and the National Construction Employers Coun-cil, is little more than a point-by-point re-sponse to Roundtable recommendations. The two hundred chief executive officers have accomplished many of their goals. They have succeeded in frightening many union leaders, muting adversarial relations, and setting the tone for the present-day agenda in construction labor relations. "The change in attitude," claims Daniel Kuise, director of collective bargaining services for the AGC, "has been phenomenal."[24]

16

The Rise of the Open Shop

Just as he had the year before, John Burns, secretary of the Massachusetts State Council of Carpenters, told delegates to the 1970 convention that construction employment in the state had reached an all-time high. Burns noted, however, that opportunities were growing in the nonunion sector as well. The Associated Builders and Contractors (ABC), a national organization of open shop builders, had chartered a Massachusetts chapter in 1968. At the time, slightly over forty members had signed on. Two years later, the chapter had 150 members holding millions of dollars worth of new construction contracts. Executive Board member Thomas Moseley seconded Burns's alarm. Residential work in the Brockton and Norfolk area was built entirely nonunion, he reported, and these once small housebuilders in the ABC were "now in the bracket where they are taking on projects amounting to a million dollars or more."[1]

The industry was in a whirlwind of change. After two decades of virtually uninterrupted record revenues for contractors and escalating wages for workers in a comfortably unionized environment, the operating assumptions were suddenly up for grabs. Public officials, pre-viously content to criticize the industry from the sidelines, were being pressed by the Business Roundtable and other "reformers" to intervene in the industry. With rare exceptions, federal policy had not ventured beyond jawboning in construction labor relations. The major experiment of the Johnson administration—Labor Secretary Willard Wirtz's suggestion to exchange moderated wage increases for a guaranteed annual employment scheme—had been decisively rejected by both contractors and unions during a New Jersey strike.

In January of 1971, President Richard Nixon warned industry leaders to voluntarily slow the wage–price spiral or face governmental action. The following month, Nixon suspended the Davis–Bacon Act, a law that fixes the pay scale (on the basis of the "prevailing" wage in the area) on federally assisted construction projects. On March 29, the White House restored the legislation and announced the formation of the Construction Industry Stabilization Committee (CISC) headed by Harvard professor and construction insider John Dunlop. CISC was empowered to establish wage guidelines, police collective bargaining agreements, and return unsatisfactory settlements for further negotiation. By the time CISC was disbanded in the spring of 1974, both strikes and wage increases had been significantly reduced.

CISC's effectiveness was only partly attributable to its heavy hand. A number of developments were crystallizing in the early 1970s to make federal intervention possible. International officials of the building trades unions, stung by public criticism of their organizations and irritated by the fiercely independent bargaining stances of their locals, quietly welcomed Nixon's actions as a vehicle to yank free-wheeling locals back into the centralized folds of the Internationals. Several Washington journalists reported International leadership's off-the-record glee whenever CISC slapped down an "excessive" local demand.[2]

Local officers and rank-and-file members wondered, too, if basic principles of unionism had suffered during the boom years, particularly in the arena of jurisdictional disputes. The repeated interunion conflicts had brought forth some of the harshest and most embarrassing criticism of the building trades. The president of the International Union of Operating Engineers reluctantly attributed many of the chronic jurisdictional problems to "plain thievery among the trades."[3] The 1960s had witnessed a near tripling of strikes over work assignments. The combined construction unions fought with nonconstruction labor unions, such as United Mine Workers District 50, to maintain exclusive control over the industry. The building trades' inward focus had long alienated them from the rest of the labor movement. In 1968, for example, less than one-quarter of the carpenters' local unions in Massachusetts bothered to affiliate with the state AFL–CIO.

Further, the trades were embroiled in a never-ending series of battles with one another. The carpenters fought with the laborers over the erection of scaffolding and the stripping of concrete forms; with the lathers over the installation of drywall; with the ironworkers over setting precast concrete; and with the sheet metal workers over metal trim. Other crafts were equally guilty. Individual locals within the same trade haggled over geographical jurisdiction, as in the long-standing feud between the Springfield and Holyoke carpenters locals over the town of Chicopee. Finally, the boom years produced an explosion of bickering and political maneuvering between individuals, locals, and regions. The bulging treasuries and the growing power and prestige of the unions made the prize of union office considerably more attractive and competitive than it had once been. Every potential internal sore point festered into open conflict. The traditional rivalry between eastern and western Massachusetts managed to find its way to the floor of the annual state convention when, as Angelo Bruno relates, "we were told that anything west of Worcester was nothing but dirt and chicken farmers."

In 1964 Charles Johnson, General Executive Board member, delivered a candid and farseeing speech to Massachusetts carpenters:

We of the Building Trades always felt that we were so powerful nobody could lick us, so we decided we could fight amongst ourselves, and nobody would pay any attention to us. We were so well entrenched with our union builders and with our union contractors' associations that we thought they would take our guff forever. In different parts of the U.S. they are just turning their backs now on the craft union in this country that built America.[4]

Four years later, Director of East Coast Organizing Abe Saul echoed Johnson's scolding: "The trades are too busy squabbling with one another, too busy looking for power within the Building Trades Council to concentrate on the common enemy, the 'scab' contractor, and he has taken advantage of it."[5]

The once tiny ABC, originally little more than a casual federation of small builders, had gradually picked up members and contracts through the 1960s. At the end of the decade, union construction workers erected almost 80 percent of all new building nationally, but the pendulum was swinging the other way. By 1972, national membership in the ABC had climbed to forty-six hundred, including seventeen of the Top 400 contracting firms. Buck Mickel, president of giant Daniel Construction, indicated that the ABC's success depended heavily on Business Roundtable sponsorship. "Open shop territories," he said, "are being expanded principally by input of industry in insisting that projects be built open shop."[6] Roundtable members, such as Du Pont and Dow Chemical, refused to allow unionized contractors to enter bids on their capital projects. Even the AGC, once the domain of union contractors exclusively, offered a symbolic gesture of approval to the new order by electing their first open shop builder as president in 1973.

Ultimately CISC, the Roundtable, and other inflation-watchers found their greatest ally in the perilous workings of the market. The great postwar construction boom peaked in 1973 and then slid into a long and severe recession. In May of 1975, the Bureau of Labor Statistics reported that construction unemployment had jumped to 22 percent. Contractor failures were at their highest level in thirteen years. James Sprouse of the AGC claimed that "only the great depression surpasses [1975] as being the most difficult of times for the construction industry."[7]

The Northeast was particularly hard hit. When the Building and Construction Trades Department of the AFL–CIO released their own findings on construction unemployment, President Bob Georgine suggested that 27 percent was a more accurate national figure. He went on to say that "in New England, there are cities that make some of the distressing figures I have just cited sound like boom towns." In Massachusetts, total receipts for building employers actually decreased by 8 percent between 1972 and 1977, and in the worst years (1973–1975) construction employment dropped 26 percent, making it one of only four states in the nation to suffer a net loss.[8]

In 1972, ten thousand construction workers marched on the State House to protest Governor Francis Sargent's order halting a number of major projects. The resulting loss of public jobs reinforced the already slumping private market. In 1973, Joseph MacComiskey announced that the prospects for work on the North Shore were "worse than I have seen it in twelve years." John Greenland went further in his description of the Boston area: "The work situation over the last twelve-month period has been the worst in the last twenty years, and unemployment in the winter months reached an all-time high." Relief was years away. Tom Gunning, executive director of the BTEA, estimated that unemployment among area union carpenters ranged from 50 to 80 percent in the winter of 1976.[9] Carpenters fled the state in search of work, some as far as the pipelines of Alaska and the oil fields of Saudi Arabia.

The fortunes of the open shop mounted with the decline of union employment. "We have not only lost the Garden Type Apartments to non-union contractors," wrote John Burns in 1973, "but we are slowly losing Shopping Centers, Motels, and Industrial work to the contractors who we at one time called wheelbarrow contractors . . . because of this situation many of our members are dropping out of the union."[10] The nonunion sector was gaining just as quickly in other states. By 1975, over nine thousand companies had joined the ABC. The AGC held its first annual "Right-to-Manage Conference" to teach union contractors how to switch to a nonunion operation. A Wharton School of Business study estimated that total open shop construction overtook union building by the middle of the decade.[11]

Union negotiators, accused of holding the industry and the American public hostage just a few short years before, were suddenly granting major concessions in wages and working rules. The "give-back" bandwagon began to roll in 1972. Ironworkers in Florida froze their wages and Rhode Island laborers took a pay cut. Carpenters in Missouri agreed to work Saturdays as make-up time for foul-weather weekdays at regular pay instead of overtime. Over the next few years, construction workers across the country accepted wage freezes and cuts. In May of 1975, the Massachusetts Council of Construction Employers, representing twenty-six state employer groups, won a "Memorandum of Understanding and Intent" from all seventeen building trades. Established as guidelines for local negotiations, the document called for lower pay scales for residential and maintenance work and the "elimination of restrictive work practices."[12]

Despite the skidding numbers in total construction dollar volume, the "wheelbarrow contractors" were growing in strength. Joe Carabetta started out with a shovel in his hand and eventually built a $180-million business. One-time shoestring operations like Seppala & Aho of Ipswich, New Hampshire, joined the upper echelons of the building fraternity. Matty Aho, the eldest of the Ahos, had been in and out of Carpenters Local 48 in Fitchburg, while his brothers worked in the family business in the years after World War II. When Nashua, New Hampshire, started a development binge, Matty left the union for the final time and set the firm's sights on larger-scale projects in New Hampshire and Massachusetts. From 1972 to 1975, Seppala & Aho ran seventy-two sites in Massachusetts alone. Midway through 1976, the company was responsible for thirteen separate multimillion-dollar contracts.

Union members harbor a particular distaste for the contractors from Ipswich because of their almost theological antiunionism. The company has consistently rejected union contacts, arguing, in the words of Martin Seppala, that labor agreements are "unnecessary." "We

try to walk uprightly in the fear of God," wrote Seppala to Carpenters official Joseph Muka. "His demands far exceed the laws of men and differ considerably."[13] "They have their own religious organization through which they bond people together," Ellis Blomquist says of the Finnish-born contractors.

And when they really got going, they began importing people from Finland. You see, there's two groups in Finland. One is quite religious and the other is quite socialistic. It's the same over here. My father was a very strong socialist and Democrat. The socialists were always pushing union organization, fighting for decent wages and working conditions. The other group tended to stick to the Bible and the colorful old legends from Finland. Seppala and Aho tapped into that religious group and set up their own church practically, you know, very strong against smoking and drinking and unions. But in that way, they were able to hold people together well.

The construction unions in Massachusetts targeted major open shop builders like Seppala & Aho, Campanelli, Carabetta, and Abreen (whose treasurer, Philip Abrams, had been national president of the ABC). Eighteen hundred workers picketed an $8-million shopping mall in Swansea; fifteen hundred marched on an $8-million Hilton Hotel and shopping center complex in Natick; and smaller numbers walked picket lines at the Braintree Howard Johnson's Motor Lodge, publicly subsidized housing projects for the elderly in Dedham and Greenfield, and dozens of other construction sites across the state.

Months of unsuccessful picketing at the $6-million Harborlight Mall on Route 3A in Weymouth burst into violence in July of 1976. South Shore construction workers, struggling with 55 to 65 percent unemployment rates, charged onto the Seppala & Aho site, knocked down part of a cement block wall, and ripped up electrical and plumbing fixtures. An angry laborer told the *Quincy Patriot-Ledger:* "We didn't want to get the poor guys working there. It's the [supervisors] in the trailers we ought to be getting." Frightened by the confrontation, Weymouth selectmen temporarily suspended the developer's building permit. When construction resumed, however, the site was still manned by nonunion workers earning as little as $3 to $4 an hour.[14]

Occasionally, the picketing paid off. But more often than not, owners, developers, and open shop contractors rejected any negotiations with the unions. By the end of the decade, 60 to 70 percent of all building in the state was erected under nonunion conditions. Massachusetts pension fund figures for 1977 indicate that working union carpenters barely averaged thirty-three weeks of employment for the year. Those were the lucky ones. Thousands of union carpenters left the area or the trade altogether. Over the course of the decade, statewide membership in the Carpenters Union dropped by one-third.[15] The national picture was equally disquieting. The UBCJA lost 91,000 members from 1975 to 1981, and the recession in the first term of the Reagan administration only aggravated an already difficult situation. In October 1982, over 1.2 million building trades workers were out of work. The 23 percent construction unemployment rate was the highest in the thirty-four years that the Bureau of Labor Statistics had kept such records.

Fierce competition over the limited number of construction contracts gave an additional edge to open-shop budget operations with low labor costs. Most industry observers agree that nonunion contractors currently hold a 70 percent market share of the industry. That estimate includes virtually complete mastery over residential and light commercial work and a surprising portion of the industrial and large-scale public market. Open shop firms accounted for 29 percent of the *Engineering News-Record's* Top 400 in 1984. And, even more suggestive of future directions, eight of the nation's top ten contractors were "double-breasted," that is, they incorporated both union and nonunion subsidiaries.[16]

The one-two punch of open shop success and declining construction volume inevitably knocked down union wages and benefits. Freezes and rollbacks actually outnumbered increases in 1984 union contracts. "Never before has there been a reduction in wages and fringes as great," said Robert Gasperow of the Construction Labor Research Council. First-year hikes averaged just 8 cents or 0.4 percent, the lowest average in the post–World War II era. Union negotiators barely improved their standing in 1985, winning a paltry 1.6 percent average increase. The slim pickings are, in the words of AGC President Doug Pitcock, "a reflection of realism in the marketplace." The 1984 and 1985 contract negotiating sessions marked the culmination of a decade of retrenchment for unionized construction workers. Despite occasional gains in union strongholds with booming markets, the cumulative impact of the 1970s and 1980s has sent construction unions reeling and granted employers a measure of uncontested authority they had not enjoyed since the 1920s. "The industry is rightly rejoicing," crowed the *Engineering News-Record* in mid-1984, "over the course that its collective bargaining has taken."[17]

[On] any scheme that requires Capital & intelligence they come to Boston for help.
—John Murray Forbes,
nineteenth-century financier

Massachusetts has always been a divided state. From John Winthrop's "City on the Hill" to Kevin White's "World-Class City," Boston's leaders have entertained an elaborately constructed self-conception of uniqueness. Birthplace and training ground of the nation's educational and medical elite, sober-minded legal and financial wizards, cultural arbiters, and political rulers, proper Bostonians have looked east to Athens, Paris, and London for role models as cosmopolitan centers of civilization. A casual westward glance induced shudders of embarrassment at the remainder of the commonwealth with its tolerably quaint but assuredly provincial towns and villages. The political and economic divisions have been real and continue to rankle. Boston has certainly stood apart in the field of construction. The city's architectural designs have been more experimental, the buildings taller, and the wages higher. For the carpenters of Massachusetts, the chasm has never been greater than it is today.

The drought of the 1970s had ended. During that decade, Massachusetts lost over eleven thousand construction workers. By the mid-eighties, the state's economy had recovered. Jobs are now available but the rebound has not been evenly distributed. Since the Reagan recession hit manufacturing the hardest, towns dependent on the region's still viable machine-tool industry suffered the worst shocks. Boston, on the other hand, based on a growing service sector, barely felt a tremor. In fact, beginning in 1981, the city embarked on a colossal building boom, reaching $1.4 billion in 1984 in private development alone.[18] At times, it appears that every corner in downtown Boston has a tower crane lifting materials up to workers building new offices, hotels, public buildings, retail outlets, or medical and educational facilities. The skyline is unrecognizable to anyone who has left the city for more than a few years.

Nor are there any signs of slackening. The rest of the United States remains mired in sluggish building swamps, but Boston construction investment is on the rise. As Robert Ryan, former director of the Boston Redevelopment Authority, wrote in 1983: "Development completions . . . scheduled for the next four years

will surpass, in value, those of the previous eight years."[19] Neighboring Cambridge has witnessed a parallel revamping of its physical appearance. While banking, insurance, and law firms are flocking to downtown Boston's new highrises, high-tech companies and educational consultants, eager to find a home in the long shadows of MIT and Harvard, are driving up rents on office space in Cambridge.

The feverish pace of building activity has separated Boston-area unions from the experiences of most construction locals in the country. At a time when "concessions" had become a household word, eight thousand Boston carpenters rejected a 31 percent increase in wages and benefits over two years, the largest offer ever made to a Boston construction union. After a five-week strike that brought $2-billion worth of construction to a halt, carpenters settled in July 1981 for a two-year contract with a $5.60 raise. When the contract expired in 1983, officers of the Boston District Council negotiated a 13 percent further increase, boosting the total package to $18.86 in wages plus an additional $5.98 in benefits by 1987.

As a sign of the times, the Boston carpenters broke away from joint bargaining sessions with other locals in eastern Massachusetts in 1983. Though their wage scales had differed, locals as far west as Worcester, south as far as Hingham, and north as far as Haverhill, had sat for years at the same negotiating table with carpenters from Boston. The mounting gulf in the relative fortunes between the cities convinced Boston-area officials to go it alone. Without the same optimal conditions, unionists in outlying towns have not been able to win comparable gains. By 1985, carpenters in the state's least fortunate locals received almost $5 an hour less in wages and benefits than their Boston counterparts.

As Bob Bryant notes, "You get outside [Route] 128 and you got a big, big problem." Carl Bathelt, business agent for Springfield Local 108, points out that the number of members of the Western Massachusetts Contractors Association has dwindled from two dozen to eight. Holyoke, his home base, once had eight union contractors and now is reduced to two. Angelo Bruno claims that the town of Greenfield has had only two union jobs in three years. Asked in March of 1984 what the situation was in the Northampton–Greenfield area, Bruno replied quickly—"desperate." Mitchel Mroz confirms Bruno's glum outlook. Interviewed on the same day, he reported that only 20 percent of the members of Local 402 in Northampton were then working. The inevitable migration to the big city has occurred. With opportunities waiting for experienced trades workers at higher wages, carpenters from towns in a 50- to 75-mile radius are commuting daily and frequently transferring their union books to one of the four locals in the Boston District Council.

The shortage of union jobs outside Boston is not due to the depressed state of the economy. There are construction jobs, but they are not necessarily unionized. The ABC now includes close to six hundred contractors in Massachusetts. Most are in the $3 to $5 million annual volume range, though some reach up to $90 million. Brotherhood organizer Steve Flynn told the 1984 state convention that nonunion carpenters outnumbered union carpenters in Massachusetts 3½ to 1.[20] The unions have relinquished more and more jobs in the outlying areas, falling back on publicly funded work protected by prevailing wage legislation, giant industrial and commercial projects, or construction users that have strong and comfortable ties with individual union contractors. Stephen Tocco, executive director of the Yankee chapter of the ABC, believes that 70 percent of the private work in Massachusetts is built nonunion and claims that open shop contractors are beginning to make inroads in the Boston market. The open shop has swept through the western and central regions of the commonwealth and is steadily inching toward

the eastern shores. "The first line was drawn somewhere around I-495," wrote *Boston Globe* reporter Bruce Mohl in 1984. "Then the construction unions closed ranks behind Route 128. Now they are circling the wagons around Boston."[21]

The ABC has not restricted its strategies to underbidding in their efforts to woo contracts away from the union sector. Under Tocco's skillful direction, the ABC has developed an extensive political, legal, and public relations agenda. Tocco, whose op-ed columns and quotable sayings continually pop up in the Massachusetts media, has adopted the stance of the underdog, the victimized open shop builder just out for his fair share of the American dream. Characterizing building trades officials as ruthless, power-hungry monopolists, Tocco has suggested that their policies "present a mirror image of the Robber Barons unions fought so courageously and righteously many years ago." For public consumption, ABC leaders in Massachusetts insistently deny any antiunion convictions. In one op-ed piece, Tocco even praised the leadership of the United Auto Workers for their exploratory steps toward a labor–management partnership. Unfortunately, he continued, local construction officials "continue to repeat their weary rhetorical flourishes." They fail to understand, he argued, that their "archaic and useless practices" are outmoded in an era of internationally integrated economies in which competitiveness and productivity provide the only rewards.[22]

Tocco has used this reformer image to skirt accusations of antiworker bias and recruit some surprising allies. In 1984, the ABC charged that Boston's Catholic Archdiocese's firm policy of hiring union construction companies amounted to "economic injustice" and therefore contradicted the Church's own teachings in the bishops' pastoral letter on the economy. In addition, Tocco has justified lawsuits against the commonwealth's 5:1 journeyman–apprentice rule on publicly funded projects as well as legislation to reduce the state prevailing wage rate as efforts to provide opportunities for minority workers in the face of the unions' history of racial discrimination. Despite the ABC's own dismal record on minority hiring, Tocco's rhetoric has won the support of Saundra Graham, a progressive black legislator from Cambridge, and the editorial staffs of the *Boston Globe* and the *Boston Herald.*

Union leaders have not been without their own weapons. In the 1983 contract negotiations, Boston Carpenters officials won a clause aimed at double-breasted firms that may have a profound impact on the industry. The practice of double-breasting has become widespread since the early 1970s. Almost every major contractor who has not shed his contractual ties to the unions altogether has established a nonunion alter ego. In eastern Massachusetts, the Macomber Company has Erland, Marshall has Algonquin, J. J. Walsh has Walsh Management, and on and on. A series of court rulings in the mid-seventies provided legal sanction for this procedure, requiring only a "maximum separation of operations" between the two entities. In practice, this technical nicety is often overlooked. Nonunion counterparts of union companies are seldom more than dummy corporations in a relative's name. Office space, equipment, and even payrolls overlap, as company accountants juggle both sets of books to maximize earnings. As firms gradually shift the bulk of their business from the union to the nonunion side, they offer their union employees the no-choice option of dropping their union books or unemployment. Contempt for legally mandated separation has reached such heights, says Bob Marshall, that double-breasted outfits slip magnetic signs with the nonunion name over the union firm's logo on company trucks as they cruise from one job site to the next.

Double-breasting has made a mockery of labor agreements in the industry. Contractors abide by union contracts only when they feel they must. Whenever possible, they operate in their nonunion guise, ignoring the very wage rates, fringe benefits, work rules, and safety provisions that they agreed to at the bargaining table. The provision in the Boston settlement was intended to halt the double-breasted dodge by requiring all subsidiaries of any contract signatory to abide by the terms of the agreement. This powder keg of a clause almost precipitated a strike. Employer negotiators branded the demand "un-American and communistic," but apparently the thought of jeopardizing billions of dollars of present and future development during the booming summer of 1983 overrode their philosophical objections.

The impact of this "work-preservation" clause is still not clear. While some predominantly union contractors divested their marginal nonunion companies, the worst offenders have stonewalled, refusing to act or divulge pertinent information. In March 1985, the Boston Carpenters and the Bricklayers (who have a similar clause) jointly filed forty-four lawsuits against builders seeking back payments of contractually obligated benefits from the double-breasted firms. An anticipated but problematic consequence of the contract language is a further separation of Boston and the outlying areas. Successful prosecution of the lawsuits may not be universally productive. "I'm afraid what will happen," sighs Bob Marshall, "is that the contractor will work his union arm in Boston and his nonunion arm outside of Boston because we'll have driven him out."

One creative method of generating union construction jobs across the commonwealth is the aggressive investment of union pension funds. For years, both union and employer trustees passively administered the joint funds, relying on the expertise of outside money managers. A series of reports in the 1970s revealed that the intentions of the funds' investors were not necessarily benign. Their vaunted professional experience had resulted in 4 to 5 percent rates of return and a disturbing pattern of investments. A 1979 Corporate Data Exchange study indicated that union pension fund investment fueled a significant portion of the nation's nonunion real estate development. Perhaps more shocking was the revelation that fund managers had purchased 13.5 percent of the common stock of Halliburton, the parent corporation of nonunion building giant Brown & Root. "They were using our money to put us out of work," concludes Local 67 Business Agent Barney Walsh.

Walsh took the lead in a campaign to regain control over investment policy in Massachusetts. He brought all the building trades unions plus the Carmens' Union into the Massachusetts Development Finance Foundation, a multimillion-dollar supertrust that handles roughly 10 percent of the various joint funds. "We are almost the same as a lending institution or bank as far as financing a project," reports Walsh, "except that our expertise is probably far better because our Foundation is made up of management people on one side and the people who build the projects on the other." The critical difference between a MDFF and a conventional bank loan is that the Foundation will only approve projects that are built entirely by union labor. Developers submit loan proposals to the trustees and each craft decides whether or not to participate and how much to invest. The MDFF has funded a $2.1-million office renovation in Cambridge, a $4.2-million office building expansion in Braintree, and a $23-million Hilton Hotel in Lowell that employed over three hundred union workers.

The Foundation is moving carefully, anxious to avoid legal entanglements with the suspicious administrators of the Employment Retirement Income Security Act (ERISA), the federal legislation that governs pension investment. The stakes in future investment directions are enormous. Walsh ultimately hopes to recruit the funds of every AFL–CIO union in Massachusetts into the "socially desirable" investment policies of the Foundation. In other states, funds have merged to sponsor low-income housing for workers or hotels that are built and staffed by union members. The concept of unions having a voice in *what* is built as well as *who* builds it has awesome implications considering the amount of money involved. Most observers estimate that by 1995 pension funds will climb to $3 trillion, representing 85 percent of the nation's total capital investment.

The sizable pension and annuity funds have found another unconventional home. The Boston District Council of Carpenters plans to open the doors of the First Union Federal Savings Bank of Boston on January 1, 1987. Located in a South Boston industrial park, this federally chartered savings and loan institution will be one of a handful of union-owned banks in the country. According to Richard Kronish, an adviser to the union on the project, returns on the two funds should generate sufficient earnings to free the bank to offer union carpenters and, potentially, other trade unionists highly competitive rates on residential mortgages and consumer loans.

The District Council is making other moves in the private housing market. Spurred by the critical shortage of low-and moderate-income housing in the Boston area and recent Church pronouncements on social and economic justice, the Boston Catholic Archdiocese approached the Carpenters Union to develop a joint plan to provide affordable homes for local residents. A series of meetings with union leaders representing electricians, painters, and plumbers, as well as city and state officials, has produced a definite commitment to establish an owner-operated prefabricated housing manufacturing plant in the Boston metropolitan area.

This scheme marks a significant break with traditional union wariness of manufactured housing, long viewed as a competitive threat to conventional on-site construction methods. In the past, union officials derided the quality of factory-produced systems in the hope that a reputation for poor workmanship would doom the successful marketing of prefabricated units. "Manufactured housing is here to stay," admits Andy Silins of the Boston District Council of Carpenters. "So we should get in on it, make the jobs union jobs, and guarantee a good product. Furthermore, I don't see any other way to address the problem of affordable housing."

Silins's willingness to recognize a societal problem beyond the confines of narrow union concerns and suggest an appropriate solution dovetails with the Archdiocese's agenda of increasing the supply of moderately priced shelter and encouraging worker participation programs. As of mid-1986, financial arrangements for the facility had not been finalized. A number of proposals have been floated, ranging from the buy-out of an existing company to the creation of an entirely new entity. But as Silins notes, "However it is originally financed, everyone involved wants to see it become worker owned." Assuming the project proceeds as planned, workers at the new plant will be governed by standard industrial contracts of their respective trade unions, but they will also be the standard-bearers for an often neglected tradition, that is, a "social unionism" in which attempts to cure societywide ills are made under the banners of the craft unions.

The unions have beefed up their apprenticeship programs in order to preserve their claim to represent the most highly qualified carpenters in the industry. Since 1966, union contracts have stipulated that a certain portion of the paycheck be funneled into an apprenticeship fund. The programs used to function twice a week on a local basis. Central and eastern Massachusetts carpenter apprentices now learn the craft with a new wrinkle. The $4-million Marshall Carpenters Training Center in Millbury, complete with classrooms, dormitory, cafeteria, and gym, opened its doors in

1985. Today's apprentices leave their jobs for two weeks at a time to live at the center and receive full-time instruction. It is an experimental approach, based on the assumption that one well-equipped and well-staffed central resource can provide a better education than several marginally financed programs.

Union leaders are conscious of another desperately important rehabilitation project—the image of the building trades unions. The open shop antiunion vendetta fused with the complacency of the postwar unions to draw a composite picture of construction workers as, in the words of Barney Walsh, "fat cats." Recently, union members have sought to correct the impression of an overpaid workforce by citing the litany of supportive statistics about seasonality, danger, and annual income. The contract covering Boston's carpenters from 1983 to 1987 includes a specific provision for shoring up the crumbling image. Five cents per working hour is deducted from every carpenter's paycheck and sent to the Boston Carpenters Promotional Education Program, an undertaking jointly administered by contractors and the union. The goal of the program is to promote union construction, a task currently carried out through newspaper advertisements, special supplements, highway billboards, and taxi-top signs.

Many unionists recognize that tangible contributions to their fellow citizens are the clearest expressions of concern and the best way to demonstrate a community of interests. Greenfield's carpenters have always been, according to Mitchel Mroz, "very community oriented." In the last several years, they built YMCA summer camps, buildings and dugouts for the local Little League, and undertook the extensive reconstruction of a historic covered bridge. Union carpenters across Massachusetts have been instrumental in a number of similar volunteer community projects. In the past year, Woburn and Lynn carpenters helped roof a senior citizens' center in Wilmington, Berkshire County Local 260 members built "Santa's Christmas Cave" for children in North Adams, and Boston trades workers helped rebuild a fire-gutted shelter for homeless women.

The cumulative effect of the volunteer projects, the promotional program, the Millbury center, the prefab housing plant, the bank, the pension fund investment, and the clever use of contract language has been to refocus the activities of existing union members, boost morale, create some new jobs, and remodel the unions' public face. They have not, however, arrested the open shop movement. That will happen only when the current ratio of union to nonunion trades workers is reversed. "We controlled the work when we controlled the manpower," John Flynn, chairman of the short-lived New England Construction Organizing Committee, once said. "That's all we ever had to sell." Union control over manpower implies bringing the unorganized workforce into the unions. "We have to make every carpenter a union carpenter," states Gordon Boraks. Boraks's formula may be perfect in its simplicity, but entrenched attitudes inside the construction unions make that goal considerably less simple to achieve.

"Organizing is the lifeblood of any labor organization," International officer Dick Griffin told the 1982 state convention, "[but] we don't do it." Many unionists believe that the prosperity of the postwar era rested on limiting access to the union. As long as the unions carefully regulated the total pool of skilled workers, they were able to guarantee relatively steady work. This restrictive system worked effectively until the overwhelming demand for construction labor in the 1960s outstripped the union supply and pushed builders to recruit outside the union structure. Pat Campbell, now general president of the UBCJA, underlined the connection between past union practices and current problems in 1973 when the open shop was still in the early stages of its march to domination: "A lot of the

people that we have kept out is the work force today that is competing with us."[23]

Inside the unions, the still commonplace belief that restrictive policies are superior to an open-door organizing strategy has outlived any accuracy it may have once had. But myths hang on in the absence of evidence of successful alternatives. Many union leaders have privately understood the need to expand the unions' ranks, but have been reluctant to advocate that view to the membership. After preaching the opposite for so many years, union officials fear that a call to open the doors will only instill short-term fears of increased competition for a fixed number of jobs. Staking out such a position has, in many instances, added up to political suicide for elected union officers. But some leaders, such as Griffin, believe the unions can no longer afford to ignore the need to organize.

We seem to operate as individuals and, yet, the very title of our organization is brotherhood. We seem to spend more damn time fighting with each other . . . organizing doesn't seem to be anything that we care about anymore because it requires effort and time and personal sacrifice on our part. And if we keep going like this, pretty soon we'll just have a token organization.[24]

Organizing successfully is not simply a matter of overcoming resistance among the membership. The labor movement of the 1980s is in a state of general disarray and decline. In this climate, organizing is a difficult proposition at best, and in construction it is even worse. "Organizing in the construction industry, past, present, and future, is very hard," comments Barney Walsh. "With an industrial union, you have more or less a captive audience in that building. With us, if you go to organize a job, and you hand out cards, by the time you've got enough, the people are probably laid off." In the past, the transience of the work and the workforce has prompted organizers to focus on contractors rather than workers. The customary technique involved recognizing when a budding builder was outgrowing his initial workforce of relatives, friends, and acquaintances. An organizer would then approach and explain the advantages of a contract and the union's ready supply of skilled workers that would allow the builder to expand further the scale of his operations. The tactic was simple, required little organizing exertion, and usually worked—as long as the unions controlled the labor force. However, today's large number of skilled or semiskilled nonunion craft workers makes such an appeal obsolete. Open shop contractors have demonstrated an alternate path, that is, growth through labor recruitment outside the unions.

The crisis of the last few years is now so severe that the call for full-scale organizing is once again being heard in the halls of the Internationals. The Brotherhood instituted Operation Turnaround, a campaign that Organizing Director Jim Parker says emphasizes "trying to make union contractors that we have more competitive." Parker argues that restrictive work rules must be eradicated, nonproductive contract clauses removed, and unnecessary strikes eliminated.[25] For many locals, making union contractors more "competitive" has, in practice, meant accepting wage freezes or reductions. Turnaround's orientation has dovetailed with the concessionary bargaining environment and employers' insistence that competitiveness depends on lower labor costs. Since most open shop builders set their wages in relation to the union scale, some industry analysts question the long-run effect of concessions in construction. Will they make union contractors more competitive or just drive down wages in union and nonunion sectors?

Turnaround's goal is to keep contractors currently under contract in the union fold through bargaining measures. There is also growing sentiment in the building trades to consider grass-roots organizing. Michael Lucas, national organizer for the International Brotherhood of Electrical Workers, has suggested taking "a look at what the people did who built our unions" when they recruited nonunion craftsmen "one at a time." Lucas's call poses a challenge to today's unionists. Can the carpenters of 1986 duplicate the efforts of Quincy Local 762's John Cogill who paid nightly visits to the unorganized carpenters of his community in 1912 in order to get "every man that works with carpenters' tools in the United Brotherhood?"[26] Can they match the efforts of the organizing committe of the Lawrence local that Richard Croteau recalls made successful home contacts as late as the years immediately following World War II? It is too soon to answer such questions. Bottom-up organizing has not been given much of a chance in the 1980s, but, when it is, it is certain that it will be harder now than it was forty or seventy years ago when mastery of a trade was often equated with ownership of a union card.

"We're amateurs," Griffin admits about current UBC organizing drives. "We don't know how the hell to do it."[27] As with any new direction, mistakes are inevitably made. Bob Weatherbee tells two stories of trying to organize with officers of other trades who were unprepared to take the kinds of risks that organizing in the eighties requires.

I went on a prevailing rate job with an agent from a specialty trade. He was dealing with six people from his trade's nonunion company. He said all six would have to leave, be replaced by his people, and then they could apply for the apprenticeship program in the spring.

We also approached a very large nonunion minority-owned company in Boston. The founder was antiunion, the son you could speak with. We had the whole thing rolling—carpenters and laborers. We had a foot in the door. We sat down at the table, ready to take everyone in. The laborer [agent] said the two foremen [who had been with the guy 20 years] could not be foremen. Why? "Because it's in our by-laws that you have to be a member for two years before you can be a foreman." The agent said, "The two foreman come from me." Thank you very much, the meeting is over. So you're driven into a position where you start arguing among yourselves. It's very unfortunate.

There is enough blame to be shared equally by all the trades. Some union leaders believe that coordinated organizing campaigns are the only way to overcome interunion bickering. Successful organizing, Griffin has said, "can only be accomplished through a unified program by building trades unions, not a single craft union." Joe Power points out that the declining number of workers on construction projects makes this cooperation even more crucial. "The trade structure can't last," he argues. "I was on a six-story job with two carpenters, four ironworkers, one plumber, one heating guy, and two electricians. They can break you down forever with such few numbers." The concept of industrial unionism in construction flies in the face of a century of craft unionism. Similar suggestions in the past have met with adamant opposition. But the time may have arrived, as Anthony Ramos, executive secretary-treasurer of the California State Council of Carpenters, recently suggested, for the building trades unions to "become one army instead of a bunch of battered regiments."[28]

The alternative is not promising. The inability to drop jurisdictional rivalries and approaches that were established in a more restrictive atmosphere makes any organizing almost impossible. Union officials acknowledge the current crisis, but still hold residual doubts about the wisdom of bringing more members into the union structure. Organizers who do not truly welcome new members but instead set strict rules under which newcomers may enter invariably alienate the very people they hope to organize. As a result, many un-

organized workers reject even the best organizing approaches. The unions' restrictive policies, Mike Harrington warned in 1973, are "making a bunch of union haters" out of those who were left out in the cold. Today's nonunion workers often do not believe things have changed. As one nonunion apprentice commented bitterly during an abortive organizing drive aimed at a large Massachusetts electrical contractor, "They just want the work we're doing. They don't want us."[29]

Boston's carpenters locals have each created a new full-time organizer slot in response to the open shop threat. Their main job is, according to Bob Marshall, "to hold on to what we have at the present time" rather than organize workers. They coordinate picket lines at nonunion sites and monitor open shop builders on prevailing rate jobs to ensure compliance. Their efforts have, in a number of cases, been successful in maintaining a strong union presence in Boston. The larger issue of organizing in Massachusetts and the nation as a whole has still not been addressed, however. Can the rising tide of the open shop be reversed or is it a permanent new fact of life in the construction industry? Bob Weatherbee believes that the basic nature of employer–employee relations will, at some point, dictate a need for unionism in the open shop sector. "Management will blow it themselves," he suggests. "They create their own problems. They'll push and push and push until there'll be a rebellion." Richard Croteau agrees but wonders if the unions' present course will help or hinder that inevitable development.

It will not turn around because the unions are going to make it turn around. It's going to happen the same way that it's always happened. When the people who are working on the nonunion job finally say, "Hey, what we need here is a union." That's when it will turn around.

Post-World War II tract housing. GMMA

Working Lives/Union Life:
Modern Times

174

Factory-based prefabrication has replaced many manual tasks. Here, completed wall panels roll down assembly line, ready to be shipped to the site. BPL

Doors now come from the factory pre-hung; they need only be leveled, plumbed, and nailed into the wall. UBCJA

A toppled crane. Construction ranks among the most dangerous of all occupations. BPL

Ironworker handles giant steel beam on the Prudential Tower. Work on the Prudential, from 1959 to 1965, began Boston's building boom after three decades of decline. PICA

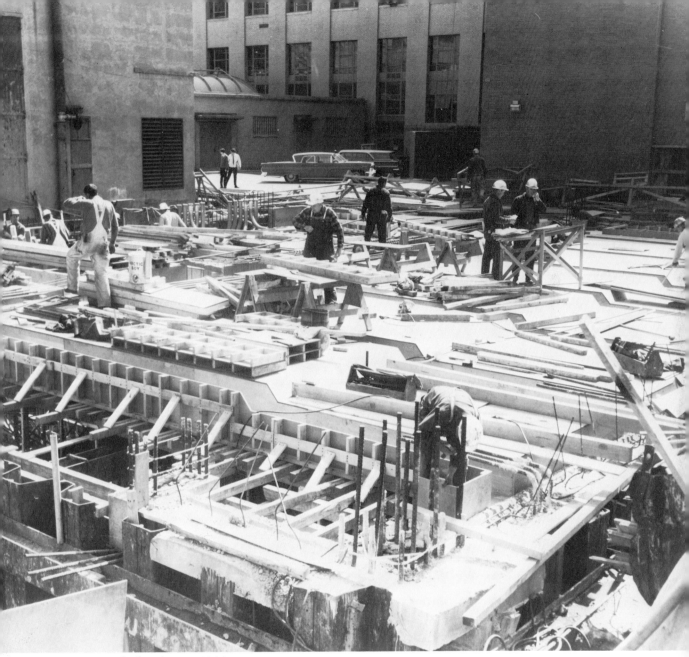

Working on a "deck" at MIT, 1966. MIT

(*Opposite page*) Building concrete forms for rerouted utility lines at MIT, 1965. MIT

179

(*Opposite page*) Laminated beams form arched church roof, c. 1975. UBCJA

(*Left*) Drywall is cut by scoring both sides of panels with utility knife. UBCJA

(*Below*) Coffee break, c. 1964. MIT

(*Left*) Pouring concrete in downtown Boston construction. BRA

(*Opposite page*) Begun in 1968 and completed in 1976, John Hancock's "glass tower" was the last major building of the 1960s boom. BRA

Pile-driving machinery at Boston's Marriott Long Wharf Hotel site, 1980. BRA

Black carpenters entered the trade in small but increasing numbers in the 1960s and 1970s. UBCJA

(*Below*) Women carpenters, marching in Washington on Solidarity Day, 1981, represent a new force in the building trades. UBCJA

(*Opposite page*) The "Cyclone" at Agawam's Riverside Park virtually complete in 1983. The 112-foot-high structure consumed over 1 million feet of lumber. CL 108

Springfield Local 108 carpenters pose in front of the "Cyclone," 1983. CL 108

17

An Industry in Transition

It is an article of faith among union carpenters that they build a vastly superior product than their nonunion counterparts. The mercurial rise of the open shop has shattered complacent notions of union invincibility, but few long-time unionists question the quality differential. "The biggest mistake that the [unionized] contractors and the union makes," asserts Ellis Blomquist, "is that they don't advertise the skill of their trained people. The contractors talk about wages all the time. They should be looking to sell the fact that they have cornered the quality workmanship market." The claim of unmatched workmanship is a chronic refrain. In his speech at the 1984 dedication of the Marshall Carpenters Training Center in Millbury, Local 33 Business Agent Robert Marshall (for whom the facility was named) cited economist Steven Allen's findings that union construction tradesmen are as much as 38 percent more productive than unorganized workers. Allen's frequently quoted study has reached mythic proportions in the building trades, serving as a scholarly affirmation of what union craftsmen have always felt in their gut.

"You can go out and see the type of work they do," insists Tom Harrington. "There's no

188

alignment. It's hit or miss." Every union carpenter has a favorite story of a botched open shop job that had to be fixed by union trades workers. Bob Weatherbee describes one situation in which a business firm accepted bids on a large door installation contract from two builders, one union and the other nonunion. The contract was awarded to the low bidder, the nonunion contractor. "The contractor went in and hung the doors. Then they went back to the union company and hired them to repair all the doors and do it all over, which cost them probably 50 percent more in the long run." Joe Power worked for years as a nonunion carpenter before joining the union in 1983. "The jobs were nightmares," he grimaces. "I worked on one just before I got into the union where, a year later, the roof still leaked and they couldn't open windows or sliding glass doors. They'd been through about a hundred guys there."

Critics say that open shop firms hire a few experienced and relatively well-paid tradesmen to supervise a dozen or more unskilled and low-waged eighteen- to twenty-five-year-olds. The ABC's Stephen Tocco does not deny this charge. On the contrary, he willingly defends it

Joe Power
By Mark Hoffman

as a proper and rational organization of an enterprise. "It doesn't bother me," he says. "That's the way it should be. You don't need to be a brain surgeon to know how to frame or put forms together. That's what American productivity and American business is all about." UBCJA Executive Board member Joseph Lia unintentionally endorses Tocco's assumptions when he discusses the impact of specialization in the industry. "You need a good layout man and a good supervisor," he said of new techniques in residential construction. "They can train somebody to do the rest. They tell somebody which side of the *X* to put the stud on, and put three nails in it. One man and 18 college kids. They get the job done."[1]

Allen's study, prepared in 1979 and revised in 1983, suggests that open shop reliance on untrained helpers and apprentices reduces quality and productivity, particularly on large commercial projects. He contends that the union advantage in productivity stems from a lower ratio of supervisors to craft workers, made possible by the better trained and self-reliant union workforce. But Allen's findings have been disputed by government and industry observers who claim that his data from the

early 1970s do not reflect the qualitative leaps in open shop managerial sophistication since that time. *Engineering News-Record*, the journal of the industry, editorialized in 1978 that "the unionized segment of the industry no longer has a monopoly over qualified workers and quality work." Tom Gunning, spokesman for an association of unionized employers in Massachusetts, agrees with *ENR*'s assessment. He warned delegates to the 1984 state carpenters' convention that union members could no longer rely on nonunion bungling to preserve their jobs: "Open shop contractors have gained in experience, have become more competent while achieving their market penetration."[2]

The question of productivity boils down to wages. If nonunion wages are low enough, open shop firms can function inefficiently with moderately skilled workers and still produce more output per labor dollar spent. On this issue there is no dispute. "Wages in nonunion house framing are abominable," says Power.

A guy I knew from before was called up last winter [1984] and asked to do frames at 93 cents a square foot. I know people down on the Cape that get $1.25. Remember, this is the head man; he pays his guys $3 and $4 an hour. That's low even by the standards of the late 60's and early 70's when I broke in. I was getting $2.50 an hour fifteen years ago.

Power's friend sits at the low end of the spectrum, but open shop construction workers do earn significantly less than unionized workers—65 nonunion cents to the union dollar, according to the Bureau of Labor Statistics.[3] Union supporters claim the differential is justified by higher productivity; open shop advocates suggest the margin explains the decline of union construction.

Whatever lower costs nonunion contractors can offer construction users flow from their organization of the labor force. "The real difference is crew makeup," says Tocco. Unencumbered by jurisdictional boundaries and restrictions on the use of apprentices and helpers (except on publicly funded projects), open shop builders pay premium wages to foremen and lead men and whatever the market will bear to everyone else. As a result, it is to the contractors' advantage to discourage the accumulation of skills by their employees in order to maintain a hierarchy of craft knowledge and corresponding pay scale. Too many accomplished mechanics would disturb the carefully constructed skill and wage pyramid. The ABC training program, known as Wheels of Learning, has been designed to match nonunion builders' limited needs. Unlike union apprenticeships that attempt to teach a wide range of skills, the Wheels of Learning is a competency system based on task. That is, if a contractor needs a framer, or a form builder, or a roofer, there is a brief training program for each aspect of the trade in which the trainee learns that particular task and no other.

The net result is a highly specialized low-paid workforce and significant cost savings for the contractor. It is a simple matter of arithmetic. A union crew of twenty-five carpenters has, at most, four or five low-paid apprentices. The remaining twenty journeymen all receive the standard hourly union rate and perform a wide variety of tasks, from the straightforward to the complex. In parallel open shop crews, the ratio is often reversed; a handful of well-paid craftsmen watch over a sizable group of novices. A Fortune 500 company recently traced the relative costs of ten years' worth of capital construction projects. They concluded that the open shop use of low-waged labor cut their construction bills substantially. The open shopper "has the freedom to manage its jobs," they reported, and does not "have to pay skilled craftsmen to do unskilled tasks."[4]

Bob Marshall knows that open shop contractors pay their workers less than union contractors but contends that they need other advantages to be truly competitive.

If they're a bona fide contractor that pays their employees a decent wage and pays all the taxes, our contractors can compete with them. It's the contractor who works out of the back of his truck, doesn't pay any unemployment tax, doesn't pay any federal or state tax, and usually hires a guy under the table that's collecting unemployment—no matter what you do, you can't compete with them.

In situations where the price of labor is fixed—publicly funded projects requiring the prevailing rate—union contractors in Massachusetts still dominate. Apparently, the knowledge and experience of both tradesmen and management in the union sector is sufficiently more extensive to maintain a hold on portions of the industry that do not allow wage-slashing. If anything, union builders hold a slight competitive edge on prevailing rate jobs that set their pay scale in accordance with the local union rate. Since approximately four dollars of the total collectively bargained package is made up of fringe benefits, union contractors need not pay payroll taxes on that deferred income. These savings represent a 5 to 7 percent advantage over open shop contractors who have no contractual benefit system with their employees and must pay the full rate in wages. Nonunion firms have made slight inroads into public sector construction in Massachusetts, but Marshall believes open shop underbidding in this market can be accomplished only through fraud.

Everyone is cheating everyone else. If you went into these companies' books, there are very few paying the taxes they're supposed to be paying or paying the employee what they're supposed to pay. They're doing it on piecework even on prevailing rate jobs—framing, shingling, roofing, siding.

Since much of the current union versus open shop debate in construction was initiated by large construction users, government policy-makers, academic consultants, and antiunion ideologues, the content has inevitably centered on cost comparisons. The perspective of the working carpenter, union and nonunion alike, has received scant attention beyond the relative size of their pay envelopes. How do the daily work lives compare in the two sectors? What of the issues that have drawn carpenters to unionization for over one hundred years? What of the problems of job security and skill accumulation in a volatile and insecure industry?

Open shop builders claim that they are able to offer more stable employment to their employees because they are not bound by rigid jurisdictional guidelines. As a result, their argument goes, if carpenters temporarily run out of work, they do not have to be laid off. They can be put to work painting, laboring, or performing simple bricklaying and electrical tasks, until the contractor arranges for more carpentry work. This flexibility may hurdle some of the industry's uncertainty on a month-to-month basis, but the overall organization of work in the open shop sector militates against long-term security. The heavy emphasis on specialization and a highly stratified workforce is an impediment to upward mobility in the trade. Since open shop builders depend on low-waged workers for a competitive advantage, there is no incentive for wage or skill advancement. On the contrary, a workforce top-heavy with skilled workers would immediately wipe out labor savings and refashion the open shop firm along conventional union lines.

Piecework as a method of payment also sacrifices long-term for short-term considerations. Piecework is a system geared for younger workers. The harder they work, the more hours they put in, the faster the job is done, the more money they earn. Pieceworkers commonly work more than eight hours a day, six or even seven days a week. Without any built-in limit on earnings, pieceworkers drive themselves

mercilessly to produce more and more. It is a system that celebrates the present and breeds contempt for the future. But the pace is ultimately impossible to sustain. After ten or fifteen years of back-breaking labor, few tradesmen have the physical stamina or the psychological desire to adhere to the system any longer. They burn out, pure and simple.

The open shop organization of work has no room for specialized workers and pieceworkers who want to trade their low-waged niches for better paying, more complex, and less physically taxing lay-out or finish tasks. There are only a handful of such slots and the cost-efficient wage pyramid cannot be sacrificed for individual employees simply because they are advancing in years. As a result, nonunion firms have a high turnover rate. Over the course of a single year, they may in fact provide more job security, but few of the workers attracted to the rugged world of nonunion construction fresh out of high school are employed in the same capacity by the original contractor five or ten years later. A small number move up the company ladder, but most leave the industry, try to make it on their own, or turn to the unions. While serving as business agent of Local 40, Bob Weatherbee has observed a steady influx of carpenters frustrated by the limited opportunities in the nonunion sector.

A lot of our membership does not come in until they're about 30 or 35 years old, until they've burnt out on house framing or forms. They know what they're doing, but the owner has canned them because some kid 22 years old has come along and will do the job for cheaper. That's when they join the union.

What is lost in many of the union–nonunion comparisons is the obvious fact that carpenters in both sectors work in the same industry. The highly dynamic state of present-day construction creates more similarities than dissimilarities for the two groups of workers. For example, whereas piecework is revered by open shop builders and banned by union work rules, the standardized character of much of today's building process tempts union contractors to violate these prohibitions whenever possible. Drywall installation on union sites is frequently rewarded on a piecework basis, a fact known to union business agents, stewards, and members who remain helpless to enforce the rule without a formal complaint from an installer.

Even more than with piecework, the general deskilling of the craft cuts across union–nonunion lines. Recognition of the impact of specialization is problematic for advocates of union construction. Veteran carpenters like Ellis Blomquist can claim that unions have cornered the quality workmanship market and, in the same breath, deplore the limited craft knowledge of the younger members of his local. These seemingly contradictory statements reflect the dual nature of a changing industry. On the one hand, a strong case can be made that a significant percentage of the commonwealth's most proficient carpenters do carry union cards in their wallets. On the other hand, subdivisions of the craft, prefabrication, and other simplifying building techniques create an on-the-job climate that reinforces highly particularized training and discourages the need for multiple skills. Michael Weinstein, who started as a union apprentice in 1972, believes there is a generation gap in competence.

If I had gotten in ten years earlier, I might be a much more skilled carpenter. Now you're typed quickly to do one kind of work. There's a real differentiation between people on the basis of how old they are, between how much they know and can do. Most of the old-timers are very well-rounded and very good carpenters. The opportunity for us to learn is just less great.

The decline in the demand for highly skilled carpenters is not a relentlessly one-way street. Gordon Boraks signed up as an apprentice ten

Michael Weinstein
By Mark Hoffman

Gordon Boraks
By Mark Hoffman

years before Weinstein. Over twenty years later, he is convinced that the finish carpentry skills he acquired in the early 1960s still stand him in good stead. Boraks contends that the call for talented carpenters varies with trends in architectural and design tastes. He points to the Habitat-induced modular craze as one extreme. "They tried to eliminate every design detail. We would just go out there with a wrench and bolt those sections together." Now he believes that the high cost of new construction combined with a renewed appreciation of historical authenticity has encouraged renovation in the Northeast and, once again, requires carpenters who can work with traditional building styles.

Boraks correctly cautions against exaggerating the demise of all skilled work, but even with the occasional step to the side, modern architects, engineers, and building materials manufacturers are traveling a clearly marked path toward simplification in construction techniques. Factory-assembled components and labor-saving machinery can be successfully adapted to remodeling as well as new work. Wood, the carpenter's most basic building block, the material that once symbolized the craft, is finding fewer and fewer applications, increasingly replaced by longer-lasting and more pliable man-made products.

The substitution of laboratory processes for organic materials has radically altered the carpenter's relationship to his or her work. Pile drivers like Tom Harrington lament the disappearance of wood in their wing of the craft. The pile drivers' attachment to lumber products was not simply nostalgic; wood provided job security. "You had the wooden piles driven first, and that was a great thing for pile drivers because God took care of that. He gave us the termites that eat the piles that you had to go ahead and replace." Today's durable piles and piers are made of poured-in-place or precast cog materials manufacturers are traveling a clearly marked path toward simplification in construction techniques. Factory-assembled components and labor-saving machinery can be successfully adapted to remodeling as well as new work. Wood, site on rollers, finally shaping the heads of the wooden timbers with an adze before driving them down to the bedrock. Now the machines move on "cats," and one worker hooks the pile up to the crane while another lets it loose at the proper location.

The phenomenon of a shrinking number of tradesmen per project is universal, from pile driving to finish work. "Jobs that would have carried a hundred people now carry twenty-five," notes Weatherbee. The introduction of "flying forms" drastically scaled down labor requirements and revolutionized concrete forming. Starting in the early and mid-1970s, carpenters no longer erected distinct forms for each wall, column, or floor. Instead they built a set of forms at the beginning of the job, attached them to sections of staging, and watched giant cranes lift the whole assembly from floor to floor as needed. The forms were set, filled with concrete, stripped after the mixture hardened, and "flown" to the floor above, ready for the next concrete pour. Bob Bryant describes the breathtaking speed of the concrete work on Turner Construction's Charles Square project in Cambridge.

The steward went to work there in July [1983] when there was nothing there. They dug the hole, came out of the hole, and went to the top on a poured-in-place job in nine months. They only employed 30 to 35 carpenters there on the concrete. You take a job like that seven or eight years ago, and they'd have 135 carpenters, and they'd be there close to two years.

The procedures at Charles Square are now the norm in concrete work. Not long ago, bridge building routinely involved crews of fifty to one hundred carpenters. Carl Bathelt reports that a recent bridge built across the Connecticut River in western Massachusetts peaked with twelve working carpenters.

Fewer workers, shorter projects. Speed, speed, and more speed. "If you don't keep up, you're let go," says Angelo Bruno. He tells a story of the foreman on the Greenfield High School site who tried to boost production by pitting one pair of carpenters against another. Bruno was the steward on the job.

The foreman points to the faster pair and says, "Look at those guys go, and the others are getting the same money." I said, "Sure, but you watch, you'll pour those forms before they pour the other ones." And, by Jesus, they did. They had to tear up the faster guys' work and start over because it was out of line. But if you're not fast enough, the contractor still lays you off. It doesn't matter if you have to go back and fix the damn thing.

Joseph Lia tells a similar story, one that rings true for every carpenter in the industry today, whether union or nonunion. He watched a man hang doors, once a task that was considered the ultimate test of a carpenter's ability. At one time, the crafty carpenter would use every trick in the book, every tool in his coveralls to shim, jockey, and nudge the door into its proper plumb, level, and square condition. The carpenter Lia observed hung a door every four minutes.

He did not square or level anything. Every screw went in with a hammer and in four minutes that door was swinging. I don't think it would last a year, so I said to him, "How can you get away with that?" He said, "That's the way they tell me to do it, that's the way I do it."[5]

———

The industry is in transition—on the site and in the contractor's office. "I remember being on jobs from beginning to end with the general contractor," Gordon Boraks says. "You'd be there to put up the fence, build the shacks, and all the way through the concrete and into the finish work. I did two years in one building." When Boraks completed his apprenticeship in the mid-1960s, it was not unusual for the majority of the total job site workforce to be on one payroll. General contractors hired carpenters, laborers, and occasionally bricklayers,

ironworkers, and painters. They employed permanent foremen and lead men for every trade and hired, laid off, and rehired entire crews as each phase of the project unfolded. The only paychecks not signed by the general contractor belonged to the mechanical trades workers—the electricians, sheet metal workers, plumbers, and steamfitters.

Specialization and an evolving construction management philosophy have deposited the broad-based general contractor into the proverbial dustbin of history. Current common wisdom compels builders to shed unwieldy payrolls and reemerge as administrators of subcontractors on the theory that specialty contractors can perform their limited functions more quickly and cheaply. *Engineering News-Record* took note of the general contractor's changing job description in a 1971 editorial: "The CM, construction manager, is a new kind of guy who, as an owner's agent, is reviewing design, estimating costs, scheduling phased construction and controlling the entire design-construct process."[6] The modern general contractor also reduces his financial risk by exchanging the uncertain outcome of the contract bid system for a guaranteed profit as a CM.

The shifting structure of the business end of the industry has cultural as well as purely economic overtones. Long-standing general contractors, such as Perini, Vappi, Volpe, Walsh Bros., O'Connell Bros., Aberthaw, and Macomber, had extensive social connections to the industry. Many of those companies' founders either rose from the ranks or had personal ties to the tradesmen and their traditions. Following the current industry practice, they have reduced their hands-on roles but still maintain significant numbers of workers in their employ. The new generation of builders tend to be graduates of business schools or the speculative world of real estate development rather than the muck of the construction site. More brokers than builders, they prefer to discuss tax shelters and money markets rather than construction labor markets. "The most significant event in the last decade has been the entrance of the MBA into the real estate–community development business," claims Anthony Ettore of the Arvida Corportion.[7] Cognizant of their meager understanding of the nuts-and-bolts of construction techniques, they rely on their subcontractors to provide the necessary expertise.

These developments coincide with the decreasing importance of labor and the increasing impact of financing on construction. The National Association of Homebuilders calculated that labor costs declined from 31 percent of the total construction price tag in 1949 to 15 percent in 1982. The share of the housing dollar going to banks, landowners, and developers, on the other hand, climbed from 31 percent to 55 percent.[8] For all the anguished outcries over excessive labor wages, the cost of money is the key to a present-day construction project. A developer's rise or fall rests as much on his ability to package financial deals and comprehend IRS regulations as on his manipulation of the labor force. During the recent building boom, Boston and Cambridge carpenters have been putting in large quantities of overtime. Area developers profit more by paying extra wages than extending interest payments. "It's cheaper to get the building up and open and ready than it is to just go along," comments Bob Weatherbee. Or as one Wall Street analyst remarked, "It used to be that the winner was the guy who had the lowest construction costs. Now the winner is the guy who is the most sophisticated in arranging financing."[9]

The changing face of management has reached down to the construction site. Job superintendents, traditionally former tradesmen, are now being recruited directly out of civil engineering programs. Turner Construction, general contractor on many of eastern Massachusetts' largest projects, is only one of many firms that rely almost exclusively on college graduates for the position of superintendent. Taking orders from young, freshly scrubbed engineers whose familiarity with construction has been confined to a textbook grates on veteran carpenters. "Years ago when you went on a job," says Weatherbee, "the super was a former carpenter. Today it's an engineer and he does not come out of the field, and he has a background of antiunion feeling. He is the guy that opens the book up and it says: 'to put up 10 feet of forms takes so many minutes.' And sometimes maybe it does, but if it's rough terrain . . . ."

The executives running modern construction firms no longer believe that field experience is a prerequisite for on-site managerial capabilities. They esteem their own professional diplomas and see no reason why that spirit of professionalism should not imbue every level of the industry. A few years back, a giant southwestern residential construction firm hired an engineer away from Procter & Gamble to become its new chief construction supervisor. "I didn't know a hammer from a 2 × 4," the man told the *Wall Street Journal*. "Fox and Jacobs asked me why I thought I could build houses and I said, 'You take men, materials, and the right equipment, and if you put them together in the right quantities you get soap. I figure you do the same to get houses.'"[10]

The unions have undergone their own process of rationalization. It has been years since even the smallest local was run out of the business agent's home. The collectively bargained funds of the postwar era—health and welfare, pension, and annuity—have brought a host of consultants, accountants, and attorneys on board. The paperwork alone ties up the better portion of a clerical work week. Union officers have also had to learn new political skills in the face of shifting industry trends. At one time, the agents only played to two constituencies— the union membership and the employers. The relationships were regular, direct, blunt, and free of outside meddling. The local union was the hub of construction's industrial relations. All activity was filtered through the business agent's office and any other industry actors, from International officers to public officials, played, at most, supporting roles.

The strings attached to the growing dollar volume of publicly funded projects oblige union officials to translate bureaucratic regulations into serviceable union policy and to enter the world of lobbyists, legislators, and state and federal agencies. Instead of waiting until ground breaking to assure a union presence, the contemporary business agent tracks projects from their planning stages. Local 67 Business Agent Barney Walsh maintains, "We have to get to the City Council in Boston, or the town meetings in Milton and Dedham, or the BRA or Public Facilities, or whoever is awarding the contract before it even gets to the Dodge Reports in order to make sure that the jobs are built in the proper manner. That's sometimes five or six years of meetings prior to a project starting."

The extensive politicking and paperwork blossomed in the last decade, according to Bob Marshall. "We used to spend our time policing the job, checking on jurisdictional problems, walking the job, and saying hi to the steward. Now, it's very unusual if I get out on the job site at all. When I started in 1975, the agent worked out of a notebook in the pocket. The office work is probably four times as much now."

Not all of the changes are so recent in origin. Fifty years of legal and political developments have altered the options of the unions and their officers. The early twentieth-century business agent referred carpenters to the job and enforced working conditions. If an employer could not be deterred from violations of the contract or work rules, it was the agent's duty to take whatever actions were necessary to insist on correct procedures. Jobs were "pulled" until the contractor sat down with the agent and mended his ways. It was, as historian David Montgomery has termed it, "the old

Sam Gompers game of 'Pull the plug; shut 'er down lads and wait until the company talks.'"[11] The system was stormy but effective. Unionists and contractors engaged in a seesaw battle for power and control over the work site that shifted favorably to one side or the other depending on economic and political conditions in the industry.

The Wagner Act of 1935 brought the era of laissez-faire labor relations to an end. The legislation set a prounion precedent for federal intervention and breathed new life into industrial unionism. But for construction unions, long accustomed to a legitimately recognized seat at the bargaining table, the National Labor Relations Act represented no Magna Carta. It was, instead, largely irrelevant. The same cannot be said of the Taft–Hartley and Landrum–Griffin amendments to the NLRA in 1947 and 1959. Suddenly, severe legal restrictions stripped union business agents of their favorite tactics. "You lost your potency," insists Oscar Pratt, "as you didn't have the authority to tie up jobs." The amended labor legislation subjected strikes, boycotts, and pickets to employer charges of unfair labor practices and the possibility of stringent monetary drains on union treasuries. Business agents, once ready to "pull the plug" on a moment's notice, now hesitated to act without consulting an attorney. "In the old days," Richard Croteau remembers fondly, "it wasn't necessary to be a lawyer in order to be a BA."

Business agents, like other trade union leaders, have developed working styles in keeping with the more regimented, bureaucratized, and legally cautious character of the modern American labor movement. The mold that produced gruff, cigar-smoking leaders such as George Meany and a host of lesser-known building trades officials has been cracked. Today's big-city business agents wear their three-piece suits almost as comfortably as the professionals they rub elbows with in their daily rounds. They recognize that the decades of smoke-filled-back-room negotiations, of arbitrary and often undemocratic leadership, and a "public be damned" attitude will not wash in an era of antiunion political sensibilities. "I recognize that our image is probably the lowest it's ever been," offered Bob Marshall in 1984. "We get $17.11 an hour and people think we're robbers." While union public relation skills still lag distantly behind the slick, glossy formulations of the open shop sector and its corporate supporters, television cameras and newspaper reporters are no longer shunned like the plague.

The transition to a qualitatively different brand of unionism is neither smooth nor easy. Union leaders are torn between the old ways and the new reality. Reluctant to abandon the business unionism that once worked so well, many just hope to tread water until the current crisis somehow passes. They fear the consequences of wholesale changes in union strategies, such as the need to create a more vibrant and democratic internal union life, to redefine their job descriptions to include grass-roots organizing, and to build political alliances outside the world of the building trades. They struggled to leave their tools behind and are now afraid to jeopardize their leadership positions by adopting experimental policies in uncertain times. But the most important truth is that construction employers have forever transformed the industry and there is no turning back. Like it or not, today's union leaders are forced to consider new approaches or hopelessly scramble to keep up with the breakneck rate of change in construction.

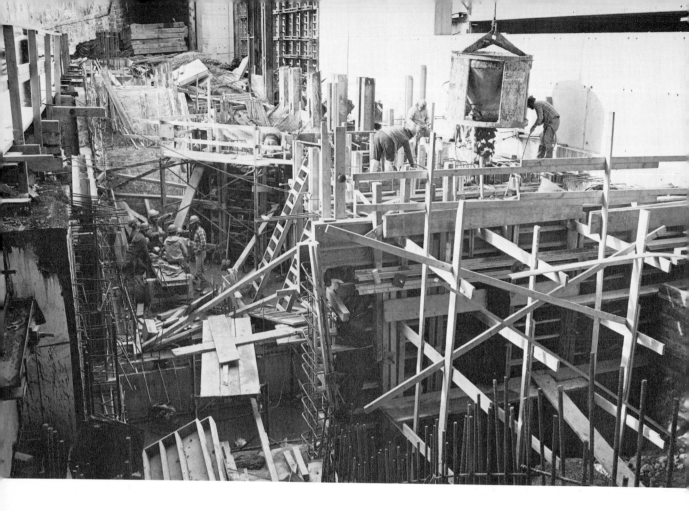

Working on the T: Photographs by John Laurenson, Jr.

John Laurenson, Jr., spent much of 1982 and 1983 picking his way through the wooden planks, concrete form panels, and reinforcing steel rods strewn over the vast Massachusetts Bay Transportation Authority construction site in the Porter Square area of Cambridge, Massachusetts. A local photographer, Laurenson hoped to capture the gritty feel of the MBTA (the "T") subway extension project that had torn up the streets of Cambridge, Somerville, and Arlington for so long.

"Dissatisfied with peripheral observation," he wrote in 1983, "I gained access to the sites in order to understand their organization and to meet and photograph their inhabitants." The results of Laurenson's underground strolls portray today's carpenter, ironworker, and laborer at work on rough concrete construction. The photos reproduced in these pages are from Laurenson's exhibit, "Working on the T: Building the Red Line Extension," displayed at the MIT Museum in 1983.

205

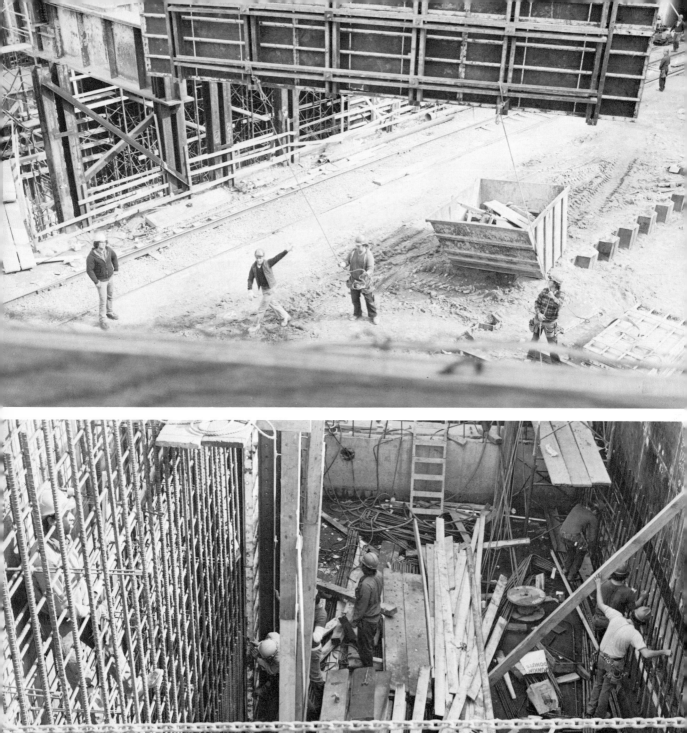

18

Knocking on the Door: Blacks and Women in Construction

When John Short was a young man, he decided he wanted to learn carpentry. He contacted a local nonunion firm. After a brief telephone conversation, the owner offered him a job. Short had been promised work as a carpenter. When he showed up, the reality fell short of his expectations. "I applied for it," says Short, "but they told me they made a mistake when they saw me."

I had to do everything. I had to dig ditches. I had to be the laborer. I had to be everything—water-boy, coffee boy. During "leisure time," I did carpentry work. When there was no laboring work to be done. They made a mistake when they saw the color of my skin. That's what it was.

Construction has always been a white indus-try. Carpenters in Massachusetts have come from a number of different backgrounds—Eng-lish, Irish, Canadian, Scandinavian, Italian, Jewish, French-Canadian, etc.—but they've usually had one thing in common, namely, the pigment of their skin. Blacks and other racial minorities have been relegated to a negligible role in the nation's building industry. Short,

who joined Local 40 in Cambridge in 1967, became a member of a highly select club. A survey conducted by the Equal Employment Opportunity Commission in the same year that Short got his card revealed that the UBCJA had a grand total of 5,284 black members, 1.6 percent of the national membership.[1]

In 1900, just 37 of the 11,500 blacks in Bos-ton were carpenters and joiners. Thirty years later, the census counted 59 black carpenters in Boston—all but 4 nonunion. Most of these men had learned their trade in the South, where there was a long and rich tradition of skilled black craftsmen, particularly carpenters, brick-layers, plasterers, and painters. When they brought their craft knowledge north, contrac-tors pulled in the welcome mats. A 1905 article on Boston's black working population found that the migrating craftsmen never had more than "intermittent occupation" and attributed this condition to the "white artisans [who] dis-like to work with them."[2]

From its inception the national Brotherhood adopted enlightened official policies on race. When a black delegate to the 1884 convention was refused service at a popular "working-man's" restaurant in Cincinnati, the body immediately declared a boycott and insisted that the establishment be held up to "public execration."[3] Despite this noble gesture and other similar statements that periodically appeared in the *Carpenter*, the national office was reluctant to impose measures to overcome racial discrimination within the organization. McGuire and other national leaders deferred to the principle of local autonomy, thereby guaranteeing Jim Crow locals in the South and a tiny handful of black union members in the North. In 1903, General President Huber condemned the existence of segregated southern locals. For many months thereafter, the pages of the union journal were filled with letter after letter from white southern carpenters attacking Huber's impudence. Though Huber did appoint a black organizer, he relented in his criticisms. He never again took up the challenge and the locals remained separate.

Most builders refused to hire nonwhite tradesmen. Sometimes they argued that blacks were not qualified; sometimes that the unions shut out racial minorities. Usually, they did not bother to justify their discriminatory policies. Joe Corbett was born in Trinidad in 1912, where, like his father, he made his living as a union carpenter. In the early 1950s, he left the West Indies, moved to Boston, and joined the Brotherhood. Despite his years of experience, membership in the Boston local did not guarantee employment.

The BA sent two of us out one morning for a job. An Englishman and myself. The contractor put the Englishman on the job and told me in no uncertain terms that, "it's a disgrace for a nigger to come and apply for a job in the shop. A nigger's rightful place is on the railroad as a porter." That's quote and unquote.

The unions were often as guilty as the contractors. Like most other whites in American society, union craftsmen had no desire to integrate blacks into their ranks. Racism was one obvious cause but additional factors operated to keep blacks out. The craft was usually passed on from one generation to the next, effectively barring anyone who had not been born with hammer in hand. What some called nepotism was standard practice in the building trades, especially before World War II. The sons and nephews of building tradesmen learned to manipulate tools as an integral part of their childhood education. Their advantages were not just a matter of better contacts inside the industry. From their earliest years, they had a leg up in craft skill on anyone else who hoped to enter the trades. As Seaton Wesley Manning observed in his 1938 investigation of black unionists in Boston, "The craft unions became more or less exclusive societies with limited memberships. This policy of exclusion was not in all instances aimed directly at the Negro though he naturally suffered from it."[4]

The general problem of exclusion has long haunted the building trades unions. There are insiders and outsiders, and the distinction rests largely on relative appreciation and respect for the idiosyncrasies of construction. In terms of peers, carpenters identify with fellow carpenters first and other building tradesmen second. Beyond that, there is a large gulf. Union carpenters have, at different times and to varying degrees, felt a bond with organized workers outside the building industry and, infrequently, with unorganized members of the workforce. The feeling of belonging to a unique world fosters a social isolationism that is only increased by the cultural homogeneity of the father-son workforce.

The stress of occupational insecurity provides an objective basis for exclusive policies. Without contractual or informal provisions for job security, unionists assume the best path to self-protection lies in limiting entry into the trades. The easiest people to exclude are those who are already outsiders. And no one has been more outside the construction industry than people of color.

In 1969, John Cort tried to explain the relationship between racism and exclusion in the building trades to the Massachusetts Advisory Committee of the U.S. Commission on Civil Rights. "This is not simply a white versus black thing," he said. "There is racial discrimination, but there is also another kind of discrimination which you might say operates regardless of race, creed or color." Cort linked the insular mentality to the trauma of the Depression.

Building trades union officials, even the younger ones, suffer from a kind of nightmarish memory of the depression of the 1930's when the building industry ground almost to a complete halt and virtually everyone in the industry was unemployed. This same nightmarish memory is handed down in tales from father to son and is extremely strong in the industry. . . . They operate on the theory that as long as there is any union member who is or might be unemployed, they are not going to admit additional union members.[5]

At the time the Advisory Committee was taking testimony in Boston, blacks had become increasingly unwilling to accept the role of outsiders. The civil rights movement had rushed past its initial emphasis on legal and political discrimination to address issues of economic justice. As Michael Harrington, president of the Massachusetts State Council of Carpenters, told the 1964 state convention, "With Negro unemployment over twice the rate of white unemployment, and with half the Negroes of the country living in poverty with an income of less than $3,000 a year, it is understandable that Negroes are pushing hard to get jobs in industries and trades where none or few have been working before."[6]

The construction industry, with its almost lily-white character and racist reputation, was a natural focus. According to Chuck Turner, who has spent much of the last fifteen years advocating affirmative action in the building trades, blacks looked to opportunities in construction because of their history of underutilized skills in the carpentry and trowel trades. In addition, notes Turner, "the sites are so visible. You're able to see very clearly what's going on, especially on jobs in your own community. Also, jobs are reorganized on a regular basis. This is a real difficulty for existing members of the trades, but it means the community looks at it as an opportunity to get people in."

In 1930, Boston's UBC locals had 4 black members; four decades later, just 10 of the 5,500 members of the Boston District Council were nonwhite. Contractor records told a similar story. Turner Construction, a major builder in the Northeast, employed 248 workers on 8 Boston projects in 1969. Thirty-four were nonwhite, and two-thirds of that group were unskilled laborers. Turner's subcontractors offered even fewer opportunities, averaging less than 3 percent minority employees. Amazingly, the record of the carpenters was better than almost any other trade. A 6 percent figure of minority apprentices in the Boston carpentry program in 1969 was significantly higher than the electricians, ironworkers, or pipefitters.[7] Joe Corbett's firsthand experience confirms this statistical portrait. There were very few black carpenters, he observes. As for the other trades, "you could count them on one hand."

In the latter half of the sixties, the federal government instituted a series of local and regional projects through the Model Cities program to increase minority representation in the industry. Model Cities projects emphasized small-scale residential and rehabilitation work in targeted urban centers with a preferential hiring policy for qualified community residents. The Boston version, the Boston Urban Redevelopment Program (BURP), opened shop in 1968, hiring mixed crews of union craftsmen and black trainees with temporary union permits. Midway through one of the BURP jobs, the contractors slashed the workforce in half.

Leo Fletcher
By Nordel Gagnon

"All the black guys were laid off, and all the French-Canadians were kept," remembers Leo Fletcher, a carpenter on the job. "Somebody from out of the city, even out of the country, getting paid to do work that I could do in my community." Fletcher and a number of other black carpenters confronted the builder and threatened to disrupt the project. After a week of discussions, the men were rehired.

The BURP experience prompted the formation of the United Community Construction Workers (UCCW), an organization of minority trades workers. At the time, federal agencies were casting about for methods to introduce more nonwhite workers into the industry. Model Cities administrators had reached agreements with the Boston Building Trades Council and the Workers Defense League to train up to two hundred carpenters, painters, bricklayers, and plumbers on projects limited to four stories in height. The unions had insisted on the building size limitation as well as a strict 3 to 1 ratio of trainees to union journeymen in order to keep the program small and out of the mainstream of unionized construction. But even within these boundaries, the programs failed to produce many qualified workers. Frus-

trated by the slow pace of change, Model Cities officials went outside the union structure and offered the UCCW funding for a six-month hands-on training program for fifty-six sheetrockers and tapers in a Dorchester apartment complex.[8]

The federal commitment to the UCCW was never more than halfhearted, particularly once the group turned to direct-action tactics. In the wake of the BURP protest, Fletcher and other UCCW activists concluded that the industry would not alter its racial composition without the threat of confrontation. "They'd always say you had to be in the union to get a job and you had to get a job to get in the union," comments Omar Cannon. "In order to get to the stage of negotiations, we had to disrupt the flow of work." In 1969, the UCCW shut down a Perini job site for three days. The following year, the group demonstrated at the $11-million addition to Boston City Hospital, calling for 40 percent of the construction jobs to go to minority workers. Site demonstrations soon became the UCCW's calling card. Says Cannon:

We had to go to the point of production, stop the job, and negotiate right on the spot. We even had people go on the job, pick up tools, and start going to work without getting paid. Some guys actually got hired like that. What worked best on any job that came into this community was to iron out an agreement before they broke ground. They would promise to take a certain amount of people through the UCCW.

In the summer of 1970, the Department of Labor approved a $680,000 grant for the Boston Hometown Plan. Created jointly by the Associated General Contractors, the building trades unions, and community representatives, the plan promised to bring two thousand minority construction workers into unionized jobs over a five-year period. A tripartite committee was formed to administer the program, recruiting through community agencies and radio and newspaper advertisements. The committee divided the recruits into four categories—journeymen, advanced trainees, trainees, and apprentices—depending on their level of experience. But despite the carefully constructed format and the ample funding, the committee never resolved its internal divisions. Chuck Turner worked with the plan in his capacity as chairman of the Boston Black United Front. "At the end of the first year," he reports, "the community actors withdrew, saying that the tripartite character made the unions and the contractors work together against the community interests." The union and management representatives continued to operate the plan and were, even by Turner's admission, "fairly effective" for a time.

The UBC was starting to correct discriminatory patterns. An updated Equal Employment Opportunity Commission (EEOC) survey in 1972 indicated that national minority membership in the Brotherhood had increased to 11.4 percent, almost a ten-point jump from five years earlier.[9] But many of the other Internationals, especially the mechanical trades, showed few signs of improvement. Overall progress remained slow and varied from city to city. In Boston, the Hometown Plan lost much of its early steam without the presence and prodding of minority participants. In the long run, the plan was not successful, admits Local 67 Business Agent Barney Walsh, but not because of the unions' role. "Contractors would use an individual on a covered job," he points out, "and when the job was completed that individual was turned loose without a light at the end of the tunnel. When the federal government then set guidelines or quotas for the unions, it was more effective, because it had teeth in it."

In 1974, the Labor Department cracked down on the sluggish plan and instituted mandatory hiring goals. Through the department's Manpower Training Act, Turner, Fletcher, and others won funding for a Third World Jobs Clearing House to act as a secondary source of labor for contractors seeking to fill hiring quotas on publicly supported jobs. For the first time, public agencies began to put some bite in enforcement policies with intermittent monitoring and mild financial sanctions. But the penalties were still too few and far between to ensure compliance. "There was no reason for contractors to contact us," notes Turner, who was executive director of the Clearing House. "And they didn't. We referred twelve jobs in our first three months."

Workers associated with the Clearing House held a series of meetings in the winter of 1975–76 to discuss what further steps to take. "People were getting mad and frustrated," claimed Thomas Ng. "When you've got three hundred to four hundred workers looking for jobs and you only give us fifteen openings, you've got to expect frustration to set in. . . . I'm not talking about minorities without skills or training, but journeymen who've been working ten years or more."[10] They decided to take their protests back to the streets. The timing of the decision was fraught with danger. The industry had already drifted into its mid-seventies tailspin, and white craftsmen who had never been pleased with the prospects of significant numbers of new entries into the unions in the best of times became adamantly opposed to heightened pressure during a recession. Long-time union members had been out of work, in some cases for two years, and were desperate for employment. They were in no mood to welcome new competition for exceedingly scarce jobs. The stage was set for an explosive confrontation in Boston, already reeling from the racially divisive battles over school busing.

In the first few months of 1976, the Third World Workers' Association, a group of black, Asian, and Hispanic workers affiliated with the Clearing House, picketed and often closed construction sites around the city. Though the demonstrations were peaceful, on-site worker resentment built with each day's pay lost to a protest-induced shutdown. The actions proved successful at first. During one three-week period, minority workers won twenty-five to thirty jobs at a Barkan site on Huntington Avenue, several buildings on Warren Street, a library in Codman Square, and the Barletta pumping station project in the South End. The pumping station protest raised the ante when it ended in a scuffle between picketers and representatives of the Building Trades Council and the anti-busing South Boston Marshals. The anger of union tradesmen spilled into the streets on May 7, as two to three thousand construction workers marched on Boston's City Hall demanding police protection on job sites and City Council defunding of the Clearing House.

Mayor Kevin White responded by placing a full complement of Tactical Patrol Force officers on the rooftops and sidewalks of the Madison Park High School project in Roxbury, a site of frequent demonstrations. But the two groups of workers had already been thoroughly polarized—black "outsiders" looking in at a sea of predominantly white faces holding the jobs they hoped for and white "insiders" viewing the protestors as one more nail in a coffin of insecurity. "I've been working now for five months," said a white worker on the Madison Park project. "Before that I was loafing for a year and a half. Do you honestly think I'm going to give up *my* job to one of these fellows?" In calmer moments of reflection, the common concerns of chronic economic hardship emerged. "We're all Depression babies," Al DiRienzo of the Massachusetts Building and Construction Trades Council told a reporter at the height of the conflict. "It's not like we don't know what Leo Fletcher and the blacks are talking about." But those moments were few and far between, vastly outweighed by the passion and panic of a perceived threat to the white construction workers' livelihoods. "In the past four years I made $7,000, $7,500,

$8,000, and $7,600," said another Madison Park worker. "I got two kids to support. It ain't enough. There's lots of us been loafing for much too long. There ain't enough work for those in the union, much less for those outside."[11]

The escalating tension bore some fruit for the Third World Workers' Association. "The whole atmosphere was very negative," Chuck Turner says, "but it also created a kind of turmoil and counterpressure in the industry. Contractors began to say they'd rather not have problems and became more cooperative." Indeed, by the end of 1976, the Clearing House had made 290 placements and had begun to function as a serious referral source. But the seething tension also took its toll. Turner believes the unending confrontations "broke the spirit of the workers who had been in the demonstrations. The pressures were too great for the returns they were getting."

The Clearing House staff and supporters decided to take a different tack. They had insisted from the start that they had no fight with white construction workers but rather with the policies of the union leaders. The turnout for the May 7 rally indicated that they overestimated any fissure between leadership and rank-and-file. They turned to the Boston Jobs Coalition policy, a more inclusive strategy. As Turner says, "We saw the only thing that would save affirmative action in the trades was an alliance with white Boston residents to break the political ties that the suburban-based unions had with the City Council." Omar Cannon makes the same point.

At one time union workers were mostly Boston residents, but they had become suburbanites. The guy from Charlestown or South Boston who at one time could be referred to a job by his uncle or a friend down the street didn't have that connection any longer. They were being pushed out too. This is why we started the Boston Jobs Coalition, to put a policy together

Chuck Turner
By Nordel Gagnon

that would address everybody so it wouldn't be just minorities and you wouldn't have a conflict between white workers and minority workers.

Census data support these observations. Though most of the Boston workforce is made up of commuters, the situation is particularly extreme in construction. In 1980, almost three of every four construction jobs in the city was filled by a nonresident. The economic advances of the post–World War II era allowed urban white craftsmen to move to roomier houses and less congested neighborhoods in the suburbs. By now, barely more than one in five white construction workers still live in the central cities of Massachusetts.[12]

Though the Coalition never made deep inroads into white neighborhoods, its broader focus and moderated tone made a residency policy more palatable to Boston politicians. In 1979, Mayor White signed an executive order implementing the Coalition's program—50 percent of the jobs to residents, 25 percent to minorities, and 10 percent to women—on all city-supported construction. Four years later, the Boston City Council upgraded White's

order when it unanimously endorsed the identical ratios as part of the Boston Jobs Residency Ordinance. In December 1984, Mayor Ray Flynn appointed a seventeen-member liaison committee to monitor the ordinance.

The affirmative action campaigns have changed the industry, though not dramatically. Many white union leaders and rank-and-file workers continue to express resentment at the presence, minimal as it is, of people of color in the trades. Overall minority participation in construction in Massachusetts remains low, in part due to enduring patterns of exclusion and in part due to the relatively small nonwhite labor force in the commonwealth in general. Blacks and Hispanics made up 4.4 percent of the metropolitan Boston construction workforce in 1980, compared to 6.4 percent of the total workforce. On the other hand, from 1960 to 1980, the number of minority workers in construction increased by 144 percent as opposed to an all-industry gain of 121 percent. The residency policy has speeded this process. City figures generally indicate compliance. In 1983, 22 percent of the work hours on contracts covered by the Jobs Residency Ordinance were filled by minority trades workers; the following year, the number climbed to 30 percent. The principle of the ordinance has won widespread acceptance, but covered projects still only account for 8 percent of total construction in Boston.[13]

At the same time that the Jobs Coalition was being formed, affirmative action advocates in Boston moved to expand their constituency in other parts of the state. Clearing House satellites were established in Cambridge, Worcester, and Springfield. These fully integrated operations met little of the fierce resistance that had characterized the Boston office. In a city plagued by a decade of intense racial conflicts, every debate over quotas and hiring goals was magnified in the glaring national spotlight. The anxiety stemming from the building slump only exacerbated the preexisting volatile atmosphere in Boston. "Smaller towns have more intimate working relations," notes Turner. "There was less pressure. We didn't have the big hassles we had in Boston." In fact, many of the building trades unions in central and western Massachusetts supported the job residency concept because of the growing influence of out-of-state contractors with their own workforces.

Attitudes in Boston eventually took a turn as well. Subsiding racial hostilities in the city created a more receptive environment. The revival of building in the early 1980s eased the panic of white workers. Above all, the rapidly growing clout of the open shop wing of the industry forced many white unionists to reconsider who their friends and enemies really were. Most union members recognize that their organizations have been slow to adapt. "Any change is going to be resisted," says John Greenland, but he claims that minority workers now participate fully in the activities of the Brotherhood. Barney Walsh concurs. "They are part of our ranks, the same as any other member of our union today." Compared to some other construction unions, the UBC has been recognized as more responsive to minority concerns. In an otherwise highly critical 1983 feature on race relations in Boston's craft unions, *Boston Globe* reporter Gary McMillan described the Carpenters as having a "decent record of admitting blacks."[14]

Chuck Turner, who now serves on the Mayor's Liaison Committee, also believes that opportunities have increased. "I'd say that affirmative action for workers of color is no longer a program that white politicians on the City Council automatically take potshots at. The construction trades are voting for it because it's part of a program that says white workers will also get their share. The basic assumptions are being respected. The unions are now negotiating in a genuine process." Bob Marshall, business manager of Boston Local 33, believes the residency requirement will actually benefit the Boston unions. "The unions have the manpower available where we can supply any job in the city with 50 percent residents. We can supply them with 20 to 25 percent minorities, and we can supply the females for the job. One of the reasons people opposed this was because they were only enforcing the requirements on the big union jobs and not on the smaller jobs."

In April of 1985, hundreds of union construction workers overflowed Gardner Auditorium at the State House to attend hearings on an Associated Builders and Contractors sponsored bill. The proposed legislation would have lowered the prevailing (i.e., union) wage scale on state-funded construction projects. ABC lobbyists contended that the aim of the bill was to provide expanded opportunities, not to break the building trades unions. As in the past, the assembled workers cheered every prounion speech and hooted every ABC speaker. But the underlying issue of race was handled differently in 1985. Massachusetts Building Trades President Tom Evers and state AFL–CIO chief Arthur Osborn hammered at the ABC logic and derided open shop minority hiring policies. They contrasted the 625 minority apprentices in the building trades unions' programs in Massachusetts to the total of 18 certificates presented by the Yankee chapter of the ABC to graduating open shop apprentices.[15] Evers argued that the time had come for the building trades to shed their narrow perspectives and build links with organizations of women and minorities against the rising antiunion tide. In a symbolic demonstration of shifting orientations and alliances, hundreds of predominantly white union workers rose to their feet to applaud black State Senator Royal Bolling's comments criticizing racially exclusive union practices while, at the same time, opposing the ABC legislation.

The building boom in Boston has relieved much of the tension between workers on the job. Relative security reduces the perception of coworkers as competitors and thereby smooths conflicts, racial and otherwise. Turner thinks "there's more general acceptance" on the site. Others agree, but ingrained attitudes die slowly. Joe Corbett's experience of the 1950s is continually being replayed into the 1980s. Nazadeen Arkil describes a fairly typical event and the toll it takes on those "outsiders" who are gradually becoming "insiders."

One day after payday, I told the guys that I was buying the coffee. My boss looked at me straight in my face and said to me, "What did you do, hit the nigger pool?" And everybody went, "Whoa." The whole job just held their breaths. I turned to him and said, "Does that mean you don't consider me a nigger?" He apologized and said, "You know, I'm just so used to you," and he went through this whole thing. The whole day, everybody was more upset about it than I was. But I said to him, "That word just rolled off your tongue a little bit too easy for me."

You gotta find the gray areas between black and white without losing your own identity, because if you're black and proud on a construction job, you might be black and dead. The attitude is "we let you in here, what the hell else do you want."

Arkil represents another recent development in the world of construction. She is both black and female. She is one of the millions of American women who entered the labor force in the 1970s. Many younger women have been attracted to nontraditional work, to the blue-collar manual occupations that pay far more than ghettoized clerical and service sector jobs. Some have sought careers in the building trades, particularly as carpenters and electricians and, to a lesser extent, as plumbers, painters, bricklayers, or ironworkers. "It's hard for a woman to make the kind of money she can make in construction," suggests Sharon Jones.

Women's entry into construction received a

Nazadeen Arkil
By Mark Hoffman

big push in 1977 when the Department of Labor included women in their affirmative action guidelines for the first time. The Department issued a three-year set of hiring goals—3.1 percent in the first year, 5 percent in the second, and 6.9 percent by the third. President Jimmy Carter's Executive Order 11246 in October 1978 strengthened the Department of Labor suggestions by mandating the quotas on any federally funded construction project over $10,000. In many ways, women were starting below the ground floor. In 1978, there were only 114 women in the entire construction labor force of Massachusetts and most of them were not craft workers.[16]

A spate of programs emerged to introduce women to building skills. The YWCA in downtown Boston offered the Co-ed Building Trades and the Non-Traditional Jobs for Women courses, but training programs really took off with the Massachusetts BTC-sponsored Women in Construction Project. Funded by the Comprehensive Employment and Training Act (CETA), sites were set up in Brighton, Carver, Lowell, and Northampton to train women in the basics of the trades and to prepare them for their journey into the construction world.

Mary Ann Williams
By Nordel Gagnon

Sharon Jones
By Mark Hoffman

saw a poster with a lot of hands holding tools in the air, and it seemed like a powerful gesture. It said, 'Women Build Your Own Future,' and I said, 'Aha, I'll build my own future.'" By the end of 1978, seventy-two students had graduated from the thirty-two-week program. Fifty-one entered union apprenticeship programs.[17]

If anything, the appearance of women on construction sites proved to be an even greater shock for white male construction workers than the presence of workers of color. The possibility that women could be equally competent craft workers threatened the rough-and-tumble "macho" image that many tradesmen had come to relish. "For a lot of these guys," Arkil says with a laugh, "their fathers, their grandfathers, and God was a carpenter. So they were born with a hammer in their hand. They think I was born with an S.O.S. in one hand and Dawn in the other." Initiation into the industry is not easy for anyone, but the transition has been particularly tough for women. Arkil says the foreman on her first job told her that she had three strikes against her—"being black, female, and left-handed." Women in the trades commonly express that, at best, they are not taken seriously and, at worst, they are abused. After finishing the Women in Construction program, Williams found her own job—a non-union site in Cambridge.

I was harrassed from the super all the way down. They told me to get my lumber out of what turned out to be the junk pile. They sent me over to dig out wood to make small footing forms. By the end of the day I had a rash all over—it was a mean little trick they played on me because that was where all the poison oak was.

Men don't have to worry about sexual harassment or being raped and women do. Some men may take it lightly, but it's a real threat. I have, and other women have, been physically grabbed. I have been assaulted by someone I worked with off my job.

Women entered the program for the earning potential and the challenge of being a pioneer in a new field for women. Mary Ann Williams had painted and loaded trucks for Jordan Marsh. She had also tried to organize a union at the department store. She explains her move into construction carpentry: "I got fed up with working there. I was riding on the train and I

There has been turnover among women in the trades, but many have persisted, finishing their apprenticeship and working steadily at the journeyman's rate. The future seems to promise more women in construction. In 1984, 235 of the construction unions' apprentices in Massachusetts were women, including 10.6 percent of the Boston Carpenters' program.[18] On-the-job acceptance of women has increased as time has passed and women have demonstrated their capacity to perform the work. "On a job that you've been on for a long time—nine months, a year—at some point, the men develop some sort of respect for you," says Jones. "OK, if not all women can do this kind of work, at least they respect that you and the other women in the program can do this kind of work."

Women report the same satisfaction in the work that men have always found. "If you're a secretary or a bank executive, you're pushing paper or answering a telephone," Faith Calhoun observes. "In construction, you put something together, you build it, and you can show something physical to someone else." Certainly, the high hourly wages are a big part of the appeal but, according to Jones, there is something more. "Most of us don't do it just for the money. We wouldn't stay in it if we didn't love it so much." Jones also places a great deal of importance on being part of the labor movement.

The women in the union really feel that their union is important to them. I don't hear that from every man I meet on the job. But the women are real committed, because they have more insights into the importance of unions, because they haven't had the opportunities to make the kind of money men do. They've seen that it's important to work in groups to get to the places they want.

For women as for racial minorities, the outsider–insider dynamic has not been finally resolved. The construction workforce has changed dramatically in the last twenty years and continues to change. The industry remains at bottom, however, a white male preserve. The changes in the labor force were initially foisted on both contractors and workers by community organizations and public agencies. With time, many of the unions switched from a stance of antagonism to one of reluctant cooperation. Many observers now wonder what the future holds with a declining commitment to the principle of affirmative action among politicians and the public at large. Outside the Boston Ordinance, enforcement of hiring quotas is rapidly scaling down. Sharon Jones is proud of her reputation as a competent union carpenter, but she worries about the impact of an adverse political climate on her future.

There's a general pervasive attitude that some day we're going to get back to the place where we don't have to have women in the trades, where the quotas aren't going to be there anymore. And they're waiting for that time and will reserve judgment until then and hopefully things will get back to the way they were. Because that happened once already, in World War II, when the Rosies were doing welding and riveting, and then the boys came back from the Army. It's happened before and they're waiting for that time to come again.

19

Who Will Build the Future?

In 1895, a civil engineer named Sanford E. Thompson was hired to develop a method of timing the movements of construction workers in order to improve their efficiency. Thompson designed and patented a "watchbook" (a thick notebook concealing two or three stopwatches) in order to perform his duties "without the knowledge of the workman who is being observed." The young engineer vehemently denied that his invention involved any deception. "There are many cases," he explained, "in which telling the workman that he is being timed in a minute way would only result in a row and in defeating the whole object of the timing."[1] After seventeen years of secretive observation, Thompson had refined his system to the point where he was able to publish his data in a massive volume entitled *Concrete Costs* (1912).

In the book, Thompson restricted his attention to the growing application of concrete as a structural material. He released the results of his studies of carpenters building concrete forms. Table 162—"Making Column, Beam, Girder, Slab, and Wall Forms on Bench Ready to Put Together"—broke down this operation into no fewer than fifty-nine separate motions. Some of the steps took as little as one or two seconds

(carrying 1 × 2 cleats 50 feet); some took as long as sixteen minutes ("average men" making 4 × 4 bolted clamps by hand). Table 163 divided the slightly more complex "Assembling and Erecting Column Forms" into seventy steps. For those who may have questioned Thompson's figures, a little note at the top of each chart assured readers that there was "no allowance made for rests and delays." This "scientifically precise" measurement of worker activities was, in Thompson's mind, a precondition to redesigning work habits more efficiently. After all, he noted, a similar time-and-motion study of bricklaying had resulted in the reduction of individual movements required to lay a single brick from eighteen to five.[2]

Thompson's original employer and mentor, and coauthor of *Concrete Costs,* was Frederick Winslow Taylor, the "father of scientific management." Taylor was the man who had elevated the stopwatch to a symbol of an all-encompassing philosophy and dedicated his life to revolutionizing American industrial management practices. Descended from a wealthy Philadelphia family, Taylor had temporarily left behind his privileged world in 1878 to enter the Midvale Steel Company workforce at the age of twenty-

219

two. Beginning as a laborer, he rapidly rose to supervisory positions, all the while studying and monitoring company machinists. On-site education completed, he left the mill in 1890 to preach his managerial doctrines full-time for the next twenty-five years.

Taylor accepted labor radical "Big Bill" Haywood's formulation that "the manager's brains are under the workman's cap" in turn-of-the-century American shops and factories. Unlike Haywood, Taylor found that condition intolerable. He believed that excessive worker craft knowledge undermined the efficient operation of an enterprise. For industrial capitalism to perform most effectively, he argued, managers needed to have a monopoly on managerial functions. The route to complete control lay in removing all decision-making responsibility from workers and reducing their role to unthinking agents of company-defined rules and laws. In Taylor's model factory, managers designed the work and delivered the orders; workers numbly carried them out. Taylorism was, according to Harry Braverman, nothing less than "a means for management to achieve control of the actual mode of performance of every labor activity, from the simplest to the most complicated."[3]

Thompson was Taylor's choice to extend these theories to the world of the building trades. *Concrete Costs* was meant to be just the first in an extended series on construction, but Taylor's death in 1915 cut the joint project short. The premises of the book they did finish, however, indicate what they had planned. The authors relentlessly criticized inept contractor practices. Labor cost estimating was guesswork, they charged, and foremen did not manage. They proposed dividing supervisory duties into seven separate positions: job designer, materials routing clerk, cost and time clerk, inspector, repair boss, instruction card man, and gang boss or foreman proper. The final two roles were the critical ones. Workers needed detailed "instruction cards," which told them not only what to do, but how to do it. "The workmen are relieved of their duties of laying out their work," they wrote, "and of deciding just how they shall do it."[4]

For builders anxious to convert to Taylorism, the authors emphasized the importance of separating workers during the transition period. "If more than one man or one gang is started at once, there is apt to be trouble," they warned. "They will talk the matter over and figure out grievances, instead of going at the job in earnest, and may refuse to work." If physical separation was impossible, Taylor and Thompson recommended firing any workers who were "holding back the rest of the gang." Once harmony had been established, an employer was ready to institute the Taylor utopia of building.

The carpenters should be divided into small gangs, usually consisting of 1 or 2 men each. Each gang should repeat the same work over and over. They may: lay out the work; make one kind of form unit repeatedly; set columns; brace columns; set posts; set girders; attach end of girders to columns; set beams; attach end of beams; and so on.[5]

Taylor believed that the centerpiece of scientific management was the definition of the "task." Once each fragment of a worker's activity had been minutely examined, dissected, and recorded, the brains under the workmen's caps could be discarded. That was the beauty of the time-and-motion study. Properly applied results, argued Taylor, automatically granted total authority to the manager. He advocated wage systems that meshed with his overall aproach. Hourly pay clearly provided little incentive for increased production. The old system of piece rates, negotiated between worker and employer, left too much power over pace in the hands of the workers. Taylor therefore called for payment based on the task. That is, foremen were advised to determine how long each particular task should take and fix a price accordingly, with no worker input.

Taylor often claimed that he did not oppose the concept of unionism, but most early twentieth-century American trade unionists clearly understood the implications of scientific management. For a number of years before, during, and after World War I, a virtual state of war existed between industrial workers and the invading hordes of stopwatch-toting Taylor disciples. Scientific management undermined the heart of worker self-respect and the ability to wield power at the workplace. As John Frey, editor of the *International Molders' Union Journal*, told a commission investigating the application of scientific management, "The greatest blow that could be delivered against unionism and the organized workers would be the separation of craft knowledge from craft skill." Frey went on,

The really essential element in [workers' craftsmanship] is not manual skill and dexterity, but something stored up in the mind of the worker. This something is partly the intimate knowledge of the character and usage of the tools, materials, and processes of the craft which . . . has enabled the workers to organize and force better terms from the employers.[6]

The pay system Taylor proposed contradicted one of the fundamental principles of building trades unionism. Among the most important trade evils that had given birth to the UBCJA were piecework, lumping, and any other method of payment that rewarded carpenters on the basis of tasks performed or individual worker competency. The Brotherhood had formalized its opposition to such wage systems in its constitution: "We are opposed to any system of grading wages in the local unions, as we deem the same demoralizing to the trade, and a further incentive to reckless competition."[7] The installation of Taylor's obsessive specialization by task would have produced a highly stratified and divided construction labor force in which each worker would receive a distinct rate of pay. In the eyes of union carpenters, this was piecework carried to its logical extreme, the destruction of craft integrity, and a deathblow to worker influence over wage rates.

On the national and local level, Brotherhood leaders insisted on a standard hourly wage scale in all negotiations. The rate was understood to be a minimum. Employers who chose to pay more to foremen and lead men were free to do so, but they were required to pay every other carpenter a uniform amount. Taylor had always contended that scientific management benefited workers as well as employers by boosting wages through the removal of wasteful and unproductive working habits. Few craftsmen bought that argument. Union members pointed out that task-oriented pay rates forced higher output, increased physical danger, and ultimately lowered rather than raised wages since employers could unilaterally alter the rate at any time. As one student of the building trades wrote in 1912 after interviewing construction union members, "This opposition to a system of payment . . . has its origins in the fear that payment by the piece will result in the gradual reduction of the piece wage, and that in the end the fast worker will receive no more than he formerly obtained on a time wage, while the slower workman will receive considerably less."[8]

Perhaps most important, unionists objected to piece rates because they promoted self-advancement at the expense of fellow workers, undercutting bonds of unionism. For this reason, the Brotherhood resisted the notion of helpers and other semiskilled categories and stuck to an arrangement of standard-waged journeymen and regulated apprentices. In an early issue of the *Carpenter*, Peter McGuire emphasized the connection between pay systems and collective union consciousness.

Piece work has a tendency to make men selfish and to give to a few who are content to work

early and late, a chance to monopolize the trade to the exclusion of better men, who while recognizing the necessity of industry, are yet unwilling to sacrifice their physical vigor by such abject slavishness to gain.[9]

In one form or another, scientific management finally swept through the nation's mass production industries. Mechanization of traditional craft skills and the introduction of the assembly line allowed manufacturing employers the freedom to institute many of Taylor's precepts if not his entire system. Construction, however, remained largely immune to his prescriptions. Unionized areas had firmly established the uniform rate and severely limited piecework by World War I. The decentralized industry, the lack of a uniform product, the persistent demand for high levels of skill, and the ability of unions to enforce work rules prevented Taylorism from taking hold. Writing in 1930, industrial relations scholar William Haber maintained that the implementation of scientific management depended on accurate indices of productivity. The construction process and product was simply too fluid, he insisted, for reliable and constant measurements. "The nature of building operations does not permit the ready introduction of methods used in housed industries," Haber concluded.[10]

The underlying themes of modern management philosophy were never far from the surface, however. Building employers of the American Plan era used the language of scientific management to justify their open shop drives of the 1920s. That language and those themes have remained remarkably consistent up to the present day. In 1921, A. Perley Ayer of nonunion Aberthaw Construction told the Boston Chamber of Commerce hearings that, since his company was not bound by a union contract, Aberthaw averaged four lower-paid helpers for every five carpenters they employed. "It is a proposition," he testified, "of

using a man for a job who is sufficiently good for the job and no better."[11] No clearer statement of the modern open shop system has ever been articulated by an ABC spokesman. Similarly, Stephen Tocco's remarks in 1985 echo the sentiments of those who came before him. According to the Yankee chapter's executive director, "When you have a foreman in the same union as the men, you have very little management control over the job." That very union rule sparked Boston's earlier builders to undertake the 1921–22 open shop campaign. The more things change, the more they stay the same. American contractors, from the Boston Master Builders Association to Business Roundtable devotees, have continually framed their disputes with the building trades unions in terms of the fundamental right to manage.

The difference between 1985 and 1925, or even 1885, is that today's building employers have successfully broken through the industry's structural barriers to "rationalization" and "modernization." They have begun a process of redesigning the nature of work in construction that perpetually eluded their frustrated predecessors. ABC contractors have triumphantly introduced organizational measures that American Plan builders only dreamed of. Apparently, Taylor's and Thompson's efforts were not in vain. Current open shop builders have swallowed some of the trappings of scientific management hook, line, and sinker. Open shop Brown & Root, the country's sixth-largest contractor, introduced efficiency experts and time-and-motion studies to their operations in 1980. The giant Houston-based firm, with almost $4 billion in contracts in 1984, used time-lapse photography to film the erection of twenty-four concrete columns. In nationwide advertisements, they explained how their "flexible open shop policy" allowed them to "make immediate changes in work assignments, material handling, and supervision to gain maximum benefits from our research."

Experts analyzed the film, breaking down the job into a series of timed operations . . . When we asked crew members for suggestions on how to get the job done faster, they came up with eight ideas, including reducing the crew from four to three members. . . . The result: a second column, was completed in five, instead of seven, hours, indicating a potential saving of 312 manhours for the column erection operation.[12]

The reorganization of construction work is by no means restricted to the open shop sector. Specialization is rampant in the union sector. The division of labor has reached new heights. Much of the modern carpenter's job is now concentrated in interior work, since basic structural carpentry has either declined in duration with the flying-form system or disappeared altogether with the use of structural steel. Taylor's and Thompson's description of an optimal "gang" has been matched and superseded by today's large-scale framing and drywall operations. Typically, within one such contractor's crew, one carpenter will lay out exclusively, another will pin metal track to the lay-out lines on the concrete floors and ceilings, another will cut and install metal studs in the tracks, another will secure preassembled metal door and window frames, several more will attach the sheetrock, yet another will insulate the walls, and on and on. Finish crews operate in the same fashion. Their work is broken down so that different carpenters hang the doors, attach the prefabricated cabinets, fasten the hardware, etc. Whether consciously or not, today's builders have followed the lines laid down by Taylor and Thompson in 1912: "The principle is to get each man accustomed to and expert in his work, to give each man a definite thing to do, and finally to let each man feel that he must work steadily in order to keep up with the gang ahead of him or out of the way of the gang behind."[13]

Despite the similarities between the union and nonunion sectors in the 1980s, there remain important differences. Tasks have been differentiated and defined—just as Taylor proposed—throughout the industry. But each specialized union carpenter (with the exception of apprentices or under-the-table pieceworkers) receives the identical union wage. And for all the work rule concessions of the past fifteen years, union regulations governing foremen's union membership, overtime, jurisdictional boundaries, the union hiring hall, journeymen–apprentice ratios, on-site safety, and the closed shop are still intact. These and other union rules have buffered the relationship of the union carpenter and his/her employer for a century. They have stood the test of time, serving as the central battleground for control of the construction site.

Unionism in the building trades also represents something intangible that transcends the legalese of contract language. The feeling of solidarity can not be seen or touched, but it is part of the distinction between workers' experiences in the two sectors of the industry. It has served as the foundation of the union carpenter's culture of cooperation and his unchanging commitment to the principle of a uniform wage. In the 1980s, the notion of "solidarity" often has a hollow ring to it, a tired word conveniently trotted out on ceremonial occasions for rhetorical rather than substantive purposes. Nonetheless, the tattered concept still carries weight for Massachusetts' union carpenters, young and old alike. "People feel like they're in something together," observes Michael Weinstein. "People use the word 'brother' a lot and it means something. It's not a joke. It's meaningful." The Brotherhood has seen, at its best, a community of shared values. Like any community, members must hold to their common vision or step outside. "In the old days," recalls Richard Croteau, "if a carpenter worked a nonunion job, he wasn't just fined. The rest of the members looked down on him and ostracized him." Or as Leo Coulombe put it bluntly, "I've been a union member all my life. I don't want to be a scab now."

Those sentiments stand in sharp contrast to the professed ideology of the ABC. Open shop spokesmen unapologetically equate their "merit" system with an American tradition of individualism. The stratified wages presumably separate the wheat from the chaff, providing a chance for those who are aggressively ambitious to take the opportunity and run. The advancement of one is achieved at the expense of another. The cultures of cooperation and individualism have often been suspended in a state of tension in the unionized wing of construction, but in today's nonunion building environment, the two cultures have been clearly separated and one completely discarded. "Personally, I could never be in a union,"remarks Stephen Tocco candidly, "because I think I'm better than the guy next to me. And I want to get paid more for it."

Without a union, says Joseph Petitpas, "everyone would be an individual and nobody would have a voice in nothing. If you went up for a raise, the boss would tell you, 'Hey, go home. I got another guy to take your place in a minute.'" "Without the union, we'd still be back in the Dark Ages," agrees Angelo Bruno. "No question about that. As an individual you can do nothing. As a group you can fight." But Bruno insists that the absence or presence of a union is not enough. It must be the kind of union that involves, challenges, and listens to the membership. He thinks today's locals often lack the creativity and democracy that characterized unions in the past.

At union meetings, whatever you thought was up for discussion. We would all discuss it and whatever came out of it, that was it, as long as it was for the benefit of the majority of the members. Your unions are going to hell when the people who don't fall in line are called the negative ones. It isn't always the conformists, the men who say yes all the time, who are the

best unionists. Sometimes, it's the ones who ask, who want to know what's going on—they're the real unionists.

The collective strength of such a union is frequently greater than the sum of its individual members. The presence of a union has long given carpenters the sense of power necessary to act in their own behalf. Older carpenters constantly refer to the many "quickie" strikes over contract violations, such as inadequate toilet facilities or drinking water, unsafe scaffolding, inferior staging planks, the lack of electrical grounding for power tools, or the appearance of nonunion tradesmen. Most such walkouts were settled within a few hour or days. "You always had the by-laws with you," Enock Peterson stresses. "If the contractor challenged you, you had the laws right there. He had an agreement and he was going to live up to it." Carpenters knew that their unions would back them up in a fair grievance, finding them another job if they were dismissed for speaking up. This knowledge bred confidence and assertiveness. Paul Weiner reports that his father often quit sites with poor safety conditions, knowing another job would be available. "He would say, 'If I work for you, I put this staging up right.'"

Tom Harrington describes a walkout of pile drivers during the construction of the Prudential Center. The pile driving contractor was from out of town and employed building techniques that had never been tolerated in Boston, particularly in the use of one man per skid rig. According to Harrington, three was the norm and any less endangered the individual worker and other craftsmen on the project. "We couldn't get anything straightened out on the job. So one day everybody just got mad and walked off the job and left all their gear and walked to the union hall." The workers voted overwhelmingly to stay out until the issue was resolved. "It only lasted a couple of days," continues Harrington, "before we had some of the wheels in from the outfit and got things straightened out. It ended up being much better for the contractor and for the men and the job went slick as could be after that. Men worked both in safety and in harmony."

The pile drivers of Carpenters Local 56 won that harmony through decisive and collective action. But circumstances in Massachusetts have changed since the Prudential was built. Carpenters, particularly outside metropolitan Boston, no longer take their unionism for granted. A strong labor organization provided a backdrop of security for the wildcatting pile drivers in 1959. The impulse for similar militance today is always tempered by the fear of the looming open shop. The building boom of the early and mid-1980s has temporarily put such anxieties on hold. Nonetheless, "staying competitive" remains the current union byword. Every false step is seen as an open invitation to the double-breasted and nonunion contractor.

Nineteen eighty-one opened the second century of the United Brotherhood of Carpenters and Joiners of America. The future that its members face is perhaps more uncertain than at any time in the last hundred years. Working carpenters are confronted with the most extensive on-site changes since the Civil War-era shift of millwork from hand to machine manufacture. The 1980s are clearly a decade of transition. What remains to be seen is whether this period is one of the recurrent watersheds that mark a new stage in the evolutionary process of the construction industry or if we are in the midst of a fundamental break with long-established methods of building.

The gradual transition of the craft worker from fabricator to installer is gaining steam. Preassembled components, modular construction, and simplified building techniques are rapidly replacing more complex conventional construction methods. Multistory, factory-built homes can now be shipped on flat-beds to the site, complete with fixtures, hardware, and a coat of paint, requiring only electrical and water hookups to be ready for occupancy. On-site building is broken down into more simplified and minute operations. The pace of specialization and deskilling is quickening in residential and commercial construction alike, as the industry sheds its backward image.

Each new labor-saving innovation weakens carpenters' bargaining power in relation to their employers and, consequently, threatens the basis of unionism in the industry. Events outside the industry pose equally serious problems for building trades unions. The generally weakened state of the AFL–CIO, the shaky public perceptions of unionism, and the reinvigorated political and business antilabor crusades have cut into the membership rolls and placed the Brotherhood and other construction unions on the defensive. Ironically, the unions have come to depend heavily on the good graces of federal, state, and local governments—sources of aid that traditionally inspired grave doubts and mistrust. Publicly funded projects are now a bulwark of union employment, offering the protective umbrella of the Davis–Bacon Act and parallel state prevailing wage laws. But some amendments by former secretary of labor Raymond Donovan and antiunion campaigns in a number of state legislatures have jeopardized those safeguards as well.

Carpenters used the staying power of a powerful craft identity to withstand crises in the past. From the obvious gesture of the sympathy strike to the more subtle internal life of the union, carpenters and their labor organizations created a remarkable and enduring culture of solidarity. They erected a self-contained value system that rewarded clever and artful work practices, a complete familiarity with every aspect of the building process, mutual assistance and social cohesion, and a deep and unshakable pride in the value of their contributions to their communities. These beliefs generated problems as well as solutions as a

consequence of the rapid and sharp differentiation between those "inside" and those "outside" the world, but their persistence also explains the tenacious hold of the craft union
culture. Carpenters operated in a transient and
mobile industry whose very structure forced
individual workers to develop highly personal-
ized survival schemes. Nonetheless, they
never relinquished their commitment to collective and egalitarian principles, most notably
the uniform wage, the ultimate expression of
solidarity within the trade.

Some construction analysts have argued that
building in the United States can never be
rationalized along the lines of a manufacturing
industry. Each project is too singular and
American tastes too idiosyncratic, they believe, to accommodate a mass-produced, au-
tomated building product. Most observers
believe, however, that no industry can stand
forever outside the inexorable march of technological determinism, that sophisticated manipulation of preassembled components will
combine the advantages of economies of scale
with individual design choices. In any case,
they suggest, the traditional and romantic notion of the capenter, weathered hands skillfully
guiding a hand plane through a curled forest of
wood shavings, will be restricted to work on
costly custom-built houses, the renovation of
existing buildings, or, more likely, serve as a
nostalgic memento of the past.

If all-around craft knowledge stands at the
center of a carpenter's identity, what will be
the result of the systematic destruction of trade
skill? What will become of the carpenter's elaborately constructed world-view? What criteria
can the "new" refashioned, specialized, and
deskilled carpenter use to define an equally
fulfilling social role? Who will the models be
and where will they come from, if not from the
preceding generations? And finally, can the
carpenter's collective voice—the craft union—
survive in the absence of a high level of trade
skills?

Industrial union advocates have long
claimed that the craft culture represented little
more than an arrogant and self-inflating
method of separating the political interests of
trade workers from the bulk of the working
population. If, in fact, the experience of building tradesmen and women becomes more akin
to the on-the-job situation of other workers,
more broad-based forms of labor organization
may indeed become more appropriate and fruit-
ful. The common thread would then be the
coincidence of membership in identical job
categories rather than an unusual degree of
pride in that particular occupational choice.

The passing of the craft culture, should it
happen, will not come without some very serious losses. In a society in which status and
achievement are linked to material wealth,
conspicuous consumption, and educational,
professional, and leisure-activity credentials,
carpenters have proudly and unmistakably defined their own self-worth through their work.
An autonomous working-class culture has always struggled to plant roots in American life,
yet the carpenter has managed to create a supportive and functional, if insular, ethic that
rewards collective and egalitarian behavior. In
the long run, the deskilling of the contemporary carpenter seems to be a foregone conclusion. If and when that occurs, it is very possible
that not only carpenters but all of us will lose
an important alternative tradition to mainstream American values and work culture.

Notes

CHAPTER 1

1. Leighton speech in Massachusetts State Carpenters Convention Proceedings, 1969, p. 40; M. R. Lefkoe, *The Crisis in Construction: There Is an Answer* (Washington, D.C.: Bureau of National Affairs, 1970), p. 159.

2. Jeannie Attie and Allen Steinberg, *Carpenters—New York State: 100 Years of Progress* (Albion, N.Y.: New York State Council of Carpenters, 1982), p. 50.

3. Massachusetts Department of Labor and Industries, *Annual Report, 1879* (Boston: Bureau of Statistics of Labor, 1879), p. 137.

4. *Boston Globe,* February 2, 1921.

5. U. S. Bureau of the Census, *1982 Census of Construction Industries* (Washington, D. C.: Government Printing Office, 1985).

6. U. S. Department of Labor, *Annual Construction Industry Report, April 1980* (Washington, D.C.: Government Printing Office, 1980).

7. John R. Commons et al., *History of Labour in the United States,* vol. 1 (New York: Macmillan, 1918), p. 341.

8. *Carpenter,* July 1897.

CHAPTER 2

1. W. A. Starrett, *Skyscrapers: And the Men Who Build Them* (New York: Scribner's, 1928), p. 288.

2. "The Industry Capitalism Forgot," *Fortune,* August 1947.

3. U. S. Bureau of the Census, *1982 Census of Construction Industries* (Washington, D.C.: Government Printing Office, 1985).

4. *Ibid.*

5. *Ibid.*

6. Quoted in William Haber, *Industrial Relations in the Building Industry* (Cambridge: Harvard University Press, 1930), p. 137.

7. *Carpenter,* February 1903.

CHAPTER 3

1. John R. Commons et al., *A Documentary History of American Industrial Society,* vol. 6 (New York: Russell & Russell, 1958), p. 99.

2. John Winthrop, *Winthrop's Journal,* ed. James Kendall Hosmer (New York: Scribner's, 1908), vol. 1, p. 112.

3. Richard B. Morris, *Government and Labor in Early America* (New York: Columbia University Press, 1946), p. 67.

4. U.S. Department of Labor, Bureau of Apprenticeship and Training, *Apprenticeship: Past and Present* (Washington, D.C.: Government Printing Office, 1982), p. 8.

5. Quoted in Philip Foner, *History of the Labor Movement in the United States*, vol. 1 (New York: International Publishers, 1947), p. 26.

6. Walter Galenson, *The United Brotherhood of Carpenters: The First Hundred Years* (Cambridge: Harvard University Press, 1983), p. 1.

7. John R. Commons et al., *History of Labour in*

the United States, vol. 1 (New York: Macmillan, 1918), pp. 70–71.

8. Commons, *Documentary History,* vol. 6, p. 76.

9. *Ibid,* p. 79.

10. *Ibid,* pp. 83, 86; Commons, *History of Labour,* vol. 1, p. 311.

11. Commons, *Documentary History,* vol. 6, p. 97.

12. Stephan Thernstrom, *Poverty and Progress* (Cambridge: Harvard University Press, 1964), p. 14.

13. *Ibid.,* pp. 92, 93.

14. Cited in Frederick S. Deibler, "The Amalgamated Wood Workers International Union of America" (Ph. D. dissertation, University of Wisconsin, 1912), p. 35.

15. Robert A. Christie, *Empire in Wood: A History of the Carpenters' Union,* Cornell Studies in Industrial and Labor Relations, vol. 3 (Ithaca: Cornell University Press, 1957), p. 26.

16. Deibler, "Amalgamated Wood Workers," p. 26.

17. *Carpenter,* June 1906; letter to *Carpenter,* September 1899.

18. *John Swinton's Paper* (New York), February 17, 1884.

19. Massachusetts Department of Labor and Industries, *Annual Report 1869* (Boston: Bureau of Statistics of Labor, 1869).

20. *Labor Leader,* March 12, 1887.

21. See Testimony of P. J. McGuire before U. S. Industrial Commission, April 20, 1899, p. 43.

22. *Carpenter,* February 1899.

CHAPTER 4

1. *New York Tribune,* September 18, 1877; cited in Philip Foner, *History of the Labor Movement in the United States,* vol. 1 (New York: International Publishers, 1947), p. 439.

2. *Carpenter,* May 1881.

3. See Annual Handbook of the Massachusetts State Council of the United Brotherhood of Carpenters and Joiners of America (n.p., n.p., 1895); and records from International office of UBCJA, Washington, D.C.

4. Cited in Jeremy Brecher, *Strike* (Boston: South End Press, 1977), p. 40.

5. Jama Lazerow, "The Workingman's Hours: The 1886 Labor Uprising in Boston," *Labor History* 21, no. 2 (1980), p. 202.

6. *Boston Globe,* May 2, 3, 1886.

7. *Ibid.,* May 5, 8, 1886.

8. *Ibid.,* May 8, 6, 1886.

9. *Ibid.,* May 3, 4, 14, 1886.

10. *Ibid.,* May 11, 1886.

11. *Labor Leader,* June 1886; *Boston Globe,* May 18, 14, 1886.

12. *Boston Globe,* May 21, 1886.

13. *Ibid.*

14. *Ibid.*

15. *Ibid.*

16. *Ibid.,* July 14, 1890.

17. Samuel Gompers to P. J. McGuire, March 20, 1890, from Samuel Gompers Letterbooks, 1883–1924, in *American Federation of Labor Records: The Samuel Gompers Era, 1877–1937* (Sanford, N.C.: Microfilming Corporation of America, 1979).

18. UBCJA *Proceedings,* 1888.

19. The Master Builders Association of Boston, *Yearbook,* 1911, p. 74–75.

20. Massachusetts Handbook, 1895; *Boston Globe,* April 2, 4, 5, 7, 14, 15, 16, 22, 28, 30, 1890; *Springfield Daily Republican,* April 18, 1890.

21. *Boston Globe,* April 7, 15, 28, 17, 1890.

22. *Ibid.,* April 14, 15, 16, 19, 1890.

23. *Ibid.*

24. *Ibid.*

25. *Ibid.,* May 1, 2, 1890; *Carpenter,* April 1890, for strike rules.

26. *Boston Globe,* May 2, 8, 1890.

27. *Ibid.,* May 3, 5, 6, 7, 9, 10, 13, 14, 16, 19, 1890.

28. *Ibid.,* June 20, 1890.

29. Board of Arbitration, Public Document No. 40, February 1891.

30. *Boston Globe,* July 13, 16, 1890; General Executive Board Minutes, June 1890, cited in Walter Galenson, *The United Brotherhood of Carpenters: The First Hundred Years* (Cambridge: Harvard University Press, 1983), p. 60.

31. *Boston Globe,* July 23, 25, September 2, 1890.

32. *Worcester Telegram,* June 24, 1890.

33. *Ibid.,* July 29, 8, 1890.

34. *Ibid.,* July 3, June 24, 1890.

35. *Ibid.,* July 9, 16, 18, 1890.

36. *Carpenter,* August 1890.

37. *Boston Globe,* May 2, April 22, 1886.

CHAPTER 5

1. *Boston Globe*, March 21, 1984.

2. *Ibid.*, December 6, 1893; J. J. McCook, "A Tramp Census and Its Revelations," *Forum* 15 (August 1893), pp. 753–66.

3. *Boston Globe*, March 21, 1894; *Labor Leader*, April 28, 1894; Report of the Massachusetts Bureau of the Statistics of Labor, 1894, p. 118 (hereinafter cited as Mass BSL); Massachusetts Department of Labor and Industries, *Annual Report 1894* (Boston: Bureau of Statistics of Labor, 1894), p. 118.

4. Mass BSL, 1894, May 8; Proceedings of 1894 UBC Convention, pp. 18, 19, 25, 61.

5. Mass BSL, 1901, p. 159.

6. A. F. Hardwick, ed., *History of the Springfield Central Labor Union 1887–1912* (n.p., n.p., n.d.), p. 71 (available at Springfield Public Library).

7. By-laws of the Carpenters District Council, Fall River; *Springfield Republican*, September 3, 1898.

8. Mass BSL, 1901, p. 143; Massachusetts State Board of Conciliation and Arbitration, *Annual Report 1902* (Boston: Wright & Potter, 1903), pp. 54, 60, 64.

9. The Master Builders Association of Boston, *Yearbook*, 1911, p. 71.

10. Massachusetts State Board of Conciliation and Arbitration, *Annual Report 1906*; "Agreement on Independence," 1902 article in the Potts clipping file, in author's possession. John Potts was a business agent in early twentieth-century Boston who preserved many articles on labor issues. His grandson Walter Potts has given the author this file.

11. Massachusetts State Board of Conciliation and Arbitration, *Annual Report 1906.*

12. James Motley, "Apprenticeship in the Building Trades," in *Studies in American Trade Unionism*, ed. Jacob Hollander and George Barnett (New York: Henry Holt & Co., 1912), p. 285.

13. David Brody, *Steelworkers in America* (New York: Harper & Row, 1969), pp. 27–79.

14. John Garraty, "U. S. Steel Versus Labor: The Early Years,"*Labor History* 1 (Winter 1960), p. 6; Otto Eidlitz Address in 1894, from Eidlitz Papers at New York Public Library.

15. From the AGC journal in 1921, quoted in William Haber, *Industrial Relations in the Building Industry* (Cambridge: Harvard University Press, 1930), p. 57.

16. Horowitz in *American Contractor,* September 1, 1923, quoted in *ibid.*, p. 395.

17. Sayward speech to the Congress on Industrial Conciliation and Arbitration, arranged under the auspices of the Industrial Committee of the Civic Federation, Chicago, November 13-14, 1894, pp. 79–80.

18. Haber, *Industrial Relations*, p. 136; William Ham, "Employment Relations in Construction in Boston" (Ph. D. dissertation, Harvard University, 1926), p. 324.

19. Haber, *Industrial Relations*, p. 449; Testimony of P. J. McGuire before the U. S. Industrial Commission, April 20, 1899, p. 22.

20. *Fall River Daily Herald*, May 24, 25, 26, 28, 29, June 7, 25, 1900.

21. Hardwick, *History of the Springfield Central Labor Union*, pp. 68, 69, 71, 73.

22. *Springfield Republican*, May 23, 1904.

23. Hardwick, *History of the Springfield Central Labor Union*, p. 68.

24. *Springfield Republican*, May 9, 1904.

25. *Ibid.*, August 10, 1904, August 8, 9, 10, 1904.

26. *Pittsfield Journal*, April 2, 3, 7, 9, 16, 17, 18, 19, 1906.

27. *Labor News* (Worcester), December 1, 1906, May 9, 1908, June 6, 1908; Board of Arbitration Reports, 1917, p. 113.

28. *Haverhill Evening Gazette*, April 7, 1909, May 18, 1909, May 29, 1909, December 12, 1913.

29. Mass BSL, 1913, p. 5, 1917, p. 211.

30. See Mass BSL, 1906 through 1912.

31. Mass BSL, 1910, p. 184; Massachusetts State Board of Conciliation and Arbitration *Annual Report 1916*, p. 194.

32. Hardwick, *History of the Springfield Central Labor Union*, p. 73.

CHAPTER 6

1. Report of the Massachusetts Bureau of Statistics of Labor, 1905; (hereinafter cited as Mass BSL) Mass BSL, 1900, p. 151; *Lynn Evening News*, March 17, 1899; P. J. McGuire to Gabriel Edmonston, November 10, 1883, from the Edmonston Papers, Reel 1 of the *American Federation of Labor Records:*

The Samuel Gompers Era, 1877–1937 (Sanford, N.C.: Microfilming Corporation of America, 1979).

2. McGuire Testimony, U. S. Senate, Committee on Education and Labor, August 17, 1883, p. 339; *Carpenter*, February 1893.

3. *Labor Leader*, February 5, 1887, April 23, 1892; *Labor News* (Worcester), May 9, 1908, January 21, 1911; Greenfield Local 549 Minutes, September 7, 1909, September 22, 1915; Springfield District Council of Carpenters Minutes, March 30, 1908.

4. *Labor Leader*, October 26, 1891, January 13, 1894, January 20, 1894; Mass BLS, 1895.

5. *Labor Leader*, April 7, 1894, April 28, 1894, June 23, 1894, August 11, 1894, November 23, 1895, December 14, 1895; *Labor News* (Worcester), November 27, 1909.

6. Oscar Handlin, *Boston's Immigrants: 1790–1865* (Cambridge: Harvard University Press, 1941), p. 236.

7. Marcus L. Hansen, *The Mingling of the Canadian and American Peoples,* vol. 1 (New Haven: Yale University Press, 1940), pp. 121, 168, 209, 215; Ronald A. Petrin, "Ethnicity and Political Pragmatism: The French Canadians in Massachusetts, 1885–1915" (Ph. D. dissertation, Clark University, 1983), p. 58.

8. Massachusetts Department of Labor and Industries, *Annual Report 1881* (Boston: Bureau of Statistics of Labor, 1881), p. 150.

9. UBCJA *Proceedings*, 1894, p. 21.

10. Abraham H. Belitsky, "Hiring Problems in the Building Trades, with Special Reference to the Boston Area" (Ph. D. dissertation, Harvard University 1960), p. 142.

11. Frank K. Foster, *The Evolution of a Trade Unionist* (Boston: n.p., 1901), p. 54.

12. Minutes of the Springfield District Council of Carpenters, October 29, 1906, March 4, 1907, January 27, 1908, June 16, 1930.

13. *Ibid.*, August 26, 1907; By-laws of the Worcester Carpenters District Council; By-laws of Local 351.

14. By-laws of the Worcester Carpenters District Council.

15. Fred S. Hall, "Sympathetic Strikes and Sympathetic Lockouts," *Studies in History, Economics, and Public Law* 10, no. 1, 1898, p. 63.

16. Mass BSL, 1908, pp. 520–521, 1913, pp. 80–83.

17. Mass BSL, 1910, pp. 226–229, 1911, pp. 151–55, 1912, pp. 48–53, 1913, pp. 78–87; Massachusetts State Board of Conciliation and Arbitration, *Annual Report 1914* (Boston: Wright & Potter, 1915), p. 107, *1916*, p. 121.

18. *Ibid., 1914*, p. 153.

19. Charles Reilly to Samuel Gompers, January 22, 1901, quoted in Philip Foner, *History of the Labor Movement in the United States* (New York: International Publishers, 1947), vol. 5, p. 207.

20. Quoted in William Ham, "Employment Relations in the Building Industry" (Ph. D. dissertation, Harvard University, 1926), p. 138.

21. Massachusetts State Board of Conciliation and Arbitration, *Annual Report 1914*, pp. 49–57.

CHAPTER 7

1. *Carpenter*, August 1883.

2. James Lynch, "The First Walking Delegate," *American Federationist* 7 (September 1901), p. 347. Lynch was probably not really the nation's first walking delegate in the building trades. Ira Cross reports walking delegates in San Francisco as far back as the 1860s. Ira Cross, *A History of the Labor Movement in California* (Berkeley: University of California Press, 1935), p. 58.

3. *Carpenter*, June 1881.

4. UBCJA *Proceedings*, 1888, p. 19.

5. *Labor Leader*, October 29, 1887, March 24, 1888, July 13, 1889, August 23, 1890, December 20, 1890, April 23, 1892, June 18,1892.

6. Letter from John Cogill, *Carpenter*, February 1912, p. 27.

7. *Carpenter*, April 1892; Luke Grant, "The Walking Delegate," *Outlook* 84, no. 11 (November 10, 1906), p. 616.

8. Lynch, "First Walking Delegate," p. 347.

9. *Labor Leader*, March 31, 1888.

10. Grant, "Walking Delegate," p. 617.

11. Harold Seidman, *Labor Czars* (New York: Liveright Publishing, 1938), p. 14.

12. Rudyard Kipling, "The Walking Delegate," *Century* 49, no. 2 (December 1894), pp. 289–97.

13. Leroy Scott, *The Walking Delegate* (New York: Doubleday, Page & Company, 1905), p. 33.

14. *Carpenter*, February 1897; quoted in Warren Van Tine, *The Making of the Labor Bureaucrat* (Amherst: University of Massachusetts Press, 1973), p. 99.

15. *Labor Leader*, October 19, 1889.

16. From Potts clipping file, in author's possession.

17. *Ibid.*

18. *Boston Globe*, February 8, 1904.

19. Minutes of Greenfield Local 549, August 25, 1910, August 24, 1911, March 13, 1913, April 10, 1913, August 12, 1914, October 14, 1914.

20. Robert F. Hoxie, *Trade Unionism in the United States* (New York: D. Appleton & Co., 1928), pp. 183–84.

CHAPTER 8

1. *Carpenter*, May 1917, p. 4.

2. *Ibid.*, p. 18.

3. *Boston Globe*, November 8, 9, 10, 1917.

4. Robert A. Christie, *Empire in Wood: A History of the Carpenters' Union*, Cornell Studies in Industrial and Labor Relations, vol. 3 (Ithaca: Cornell University Press, 1956), p. 226.

5. *Boston Globe*, November 9, 1917.

6. *Ibid.*, November 10, 13, 1917.

7. *The Builders: The Seventy-five Year History of the Building and Construction Trades Department, AFL–CIO* (Washington, D.C.: AFL–CIO, Building and Construction Trades Department, 1983), p. 7.

8. *Ibid.*, p. 7.

9. *Boston Globe*, February 2, 1921.

10. Jeremy Brecher, *Strike* (Boston: South End Press, 1977), pp. 103, 104; *Boston Globe*, January 24, 1921.

11. *Springfield Labor Advocate*, May 16, 1919.

12. Quoted in Brecher, *Strike*, p. 116.

13. David Montgomery, *Workers' Control in America* (Cambridge: Cambridge University Press, 1979), p. 99.

14. History Committee of the General Strike Committee, *The Seattle General Strike* (Cambridge: reprinted as a Root and Branch pamphlet, 1972), p. 6.

15. *Boston Globe*, June 2, 1919, May 21, 1919.

16. *Ibid.*, May 13, 14, 1919.

17. *Ibid.*, May 15, June 13, May 26, 1919.

18. *Ibid.*, May 13, 26, June 2, 1919.

19. *Ibid.*, June 13, 28, 1919; *Carpenter*, August 1919.

CHAPTER 9

1. John F. Nason, "The House That Jack Built," *Nation*, March 1, 1922, p. 255.

2. "The Building Situation in Boston," Hearings before a committee of the Boston Chamber of Commerce (Boston: Boston Chamber of Commerce, 1921), p. 962.

3. Herbert B. Adams, ed., *History of Cooperation in the United States* (Baltimore: Johns Hopkins University Press, 1888), pp. 86–88, 299.

4. *Labor News* (Worcester), April 8, 1921.

5. *Boston Globe*, May 30, 1919.

6. *Ibid.*, May 30, 1919, May 26, 1920.

7. *Ibid.*, May 26, 1920.

8. Mary Conyngton, "Housing: Building-Trades Unions' Construction and Housing Council of Boston," *Monthly Labor Review* 14, no. 5 (May 1922), p. 164; Nason, "House That Jack Built," p. 255.

9. Conyngton, "Housing," p. 164.

10. *Ibid.*, p. 163.

11. *Engineering News-Record*, February 16, 1984, p. 47; Nason, "House That Jack Built," p. 255.

CHAPTER 10

1. William Haber, *Industrial Relations in the Building Industry* (Cambridge: Harvard University Press, 1930), p. 401.

2. *The Builders' Record*, February 28, 1921; "The Building Situation in Boston," Hearings before a committee of the Boston Chamber of Commerce (Boston: Boston Chamber of Commerce, 1921), p. 577.

3. Boston Chamber of Commerce Hearings, p. 681.

4. *Ibid.*, p. 591.

5. *Ibid.*, pp. 681, 682, 591.

6. *Ibid.*, pp. 1019, 698; *Boston Globe*, January 21, 1921.

7. Boston Chamber of Commerce Hearings, pp. 1015, 575; *The Builders' Record*, February 28, 1921.

8. Boston Chamber of Commerce Hearings, p. 877.

9. *Boston Globe*, February 2, 1921.

10. Boston Chamber of Commerce Hearings, p. 716.

11. *Boston Globe*, January 23, 1921; Boston

Chamber of Commerce Hearings, p. 679.

12. *Boston Globe*, February 3, 1921; Boston Chamber of Commerce Hearings, p. 843.

13. Massachusetts State Board of Conciliation and Arbitration, *Annual Report 1921* (Boston: Wright & Potter, 1922), pp. 110–11.

14. *Boston Globe*, January 15, 21, 31, 1921.

15. *Ibid.*, January 21, 24, 26, 29, 31, February 2, April 15, 1921.

16. Boston Chamber of Commerce Hearings, pp. 1069, 1071.

17. *Boston Globe*, February 2, 16, 24, 1921.

18. *Ibid.*, March 24, April 2, 25, May 3, 1921.

19. *Ibid.*, March 12, 21, April 8, 15, 17, May 19, 1921.

20. *The Builders' Record*, September 7, 1921; *Boston Herald* editorial quoted in same issue of *Builders' Record*.

21. *Boston Herald*, December 28, 1921.

22. *Ibid.*, July 15, 1922.

23. *Ibid.*, February 18, March 2, 15, 1923.

24. *Ibid.*, March 8, 1923.

CHAPTER 11

1. Irving Bernstein, *The Lean Years* (Boston: Houghton Mifflin, 1970), p. 97.

2. UBCJA *Proceedings*, 1928; Spargo quoted in David Brody, *Workers in Industrial America* (New York: Oxford University Press, 1980), p. 62.

3. Robert Lynd and Helen Lynd, *Middletown* (New York: Harcourt Brace Jovanovich, 1956), pp. 80–81.

4. *Labor News* (Worcester), July 4, 1924, September 3, 1926, February 11, 1927, April 15, 1927, June 3, 1927, January 27, 1928, August 23, 1929, October 4, 1929.

5. *Boston Herald*, September 10, 1926, January 10, 1928, March 23, 1928, May 28, 1928.

6. *Ibid.*, July 23, 1926; *Labor News* (Worcester), September 11, 1925.

7. William Robinson, "Fall River: A Dying Industry," *New Republic*, June 4, 1924, pp. 38–39.

8. Proceedings of the 1925 Massachusetts State Convention of Carpenters, pp. 17, 18, 19.

9. Massachusetts Department of Labor and Industries, "Unemployment of Organized Building Tradesmen in Massachusetts," August 1, 1927, Jan-

uary 3, 1928; Louis Adamic, *My America: 1928–1938* (New York: Harper & Bros., 1938), pp. 265–66.

10. Adamic, *My America*, pp. 263, 264, 274.

11. *Labor News* (Worcester), June 29, 1928.

12. *Carpenter*, September 1930; *Boston Herald*, March 20, 1933; Springfield District Council of Carpenters Minutes, May 29, 1934.

13. *Carpenter*, June 1928, p. 48.

14. Springfield District Council of Carpenters Minutes, May 31, 1934.

15. *Carpenter*, March 1930, p. 53.

16. *Labor News* (Worcester), March 18, 1932.

17. Springfield District Council of Carpenters Minutes, October 21, 1929, September 13, 1932.

18. *Ibid.*, May 29, 1930.

19. *Ibid.*, April 7, 1930.

20. *Boston Herald*, August 29, 1930.

21. *Boston Globe*, October 22, 23, 31, November 26, 1931.

22. Adamic, *My America*, p. 298; *Boston Globe*, November 26, 1931.

23. Springfield District Council of Carpenter Minutes, August 8, 1932; Greenfield Local 549 Minutes, February 4, 1932.

24. *Boston Herald*, February 28, 1933, March 20, 1933.

25. Springfield District Council of Carpenters Minutes, May 23, 1933, August 22, 1933, October 18, 1933.

26. UBCJA *Proceedings*, 1936, p. 122; *Labor News* (Worcester), September 4, 1931.

27. UBCJA *Proceedings*, 1936, pp. 109–17.

28. Springfield District Council of Carpenters Minutes, April 1, 1931.

29. *Ibid.*, February 9, 1932, March 18, 1932.

30. *Ibid.*, May 15, 1930, August 8, 1932; *Boston Herald*, February 28, 1933; *Labor News* (Worcester), December 19, 1930.

CHAPTER 12

1. Robert McElvaine, *The Great Depression* (New York: Times Books, 1984), p. 67.

2. *Ibid.*, p. 16.

3. Michael Rogin, "Voluntarism: The Political Functions of an Anti-Political Doctrine," in *The American Labor Movement*, ed. David Brody (New York: Harper & Row, 1971), p. 112.

4. Maxwell Raddock, *Portrait of an American Labor Leader: William L. Hutcheson* (New York: American Institute of Social Science, 1955), p. 245.

5. *Boston Globe*, June 15, 1933.

6. McElvaine, *The Great Depression*, p. 172.

7. *Labor News* (Worcester), February 5, 1932.

8. Springfield District Council of Carpenters Minutes, October 5, 1933, July 16, 1934.

9. *Ibid.*, March 14, 1933, January 10, 1934.

10. *Boston Herald*, July 23, 1935; *Our World* (Boston), October 9, 1935.

11. *Our World* (Boston), November 18, 1935, March 25, 1936; *Boston Globe*, January 9, 1936.

12. *Our World* (Boston), March 25, 1936.

13. *Boston Herald*, September 25, 1937.

14. UBCJA *Proceedings*, 1936, p. 364; Massachusetts AFL Proceedings, 1938, p. 70.

15. George Raiche, "The A.F.L. Labor Movement in Springfield," *Industrial Springfield* (Springfield: Springfield Central Labor Union, 1939); *Our World* (Boston), December 23, 1935.

CHAPTER 13

1. *Boston Herald*, October 27, 1940.

2. *Ibid.*, March 31, 1941.

3. U.S. Bureau of the Census, *Construction Statistics, 1915–1964* (Washington, D. C.: Government Printing Office, 1966), pp. 2, 9.

4. *Ibid.*, p. 9.

5. 1984 Massachusetts State Carpenters Convention Proceedings, p. 87.

6. Quoted in Martin Mayer, *The Builders* (New York: Norton, 1978), p. 250.

7. 1965 Massachusetts Convention Proceedings, p. 9.

CHAPTER 14

1. William Haber, *Industrial Relations in the Building Industry* (Cambridge: Harvard University Press, 1930), p. 38.

2. David Noble, *Forces of Production* (New York: Alfred A. Knopf, 1984), p. 64.

3. Springfield District Council of Carpenters Minutes, August 8, 1932.

4. *Ibid.*, June 29, 1964.

5. William Haber and Harold Levinson, *Labor Relations and Productivity in the Building Trades* (Ann Arbor: University of Michigan Press, 1956), pp. 115, 120.

6. Abraham H. Belitsky, "Hiring Problems in the Building Trades, with Special Reference to the Boston Area" (Ph.D. dissertation, Harvard University, 1960), pp. 35–36.

7. *Ibid.*, p. 25.

8. *Ibid.*, p. 200.

CHAPTER 15

1. *Boston Globe*, April 2, 1956.

2. *Christian Science Monitor* (Boston), May 5; *Boston Globe*, May 30, 1958.

3. *Christian Science Monitor* (Boston), October 13, 1952.

4. Steven E. Miller, "The Boston Irish Political Machine, 1830–1973," unpublished paper, pp. 128, 129.

5. Anthony Yudis and Robert Lenzner, "Massachusetts Economy—The Myths and the Realities," *Boston Globe*, March 19, 1973.

6. Information from fact sheet provided by Prudential Center.

7. U.S. Bureau of the Census, *Construction Statistics, 1915–64* (Washington, D.C.: Government Printing Office, 1966), p. 42.

8. U.S. Bureau of the Census, *Employment, Hours, and Earnings, States and Areas, 1939–82*, vol. 1 (Washington D.C.: Government Printing Office, January 1984), p. 358; U.S. Bureau of the Census, *Construction Statistics, 1915–64*, p. 37; U.S. Bureau of the Census, *Census of Construction Industries, 1972, Area Series* (Washington, D.C.: Government Printing Office, 1975).

9. *Engineering News-Record*, July 17, 1969, p. 7.

10. John Dunlop, "Foreword," in D. Q. Mills, *Industrial Relations and Manpower in Construction* (Cambridge: MIT Press, 1972), p. vii.

11. *Engineering News-Record*, May 8, 1969, p. 61; and Patrick McCauley, "Economic Trends in the Construction Industry, 1965–80," *Construction Review*, May/June 1981, p. 8.

12. Abraham H. Belitsky, "Hiring Problems in the Building Trades, with Special Reference to the Boston Area" (Ph.D. dissertation, Harvard University, 1960), pp. 134–35.

13. *Engineering News-Record*, April 25, 1968, p. 76.

14. *Ibid.*, June 2, 1966, p. 58, November 24, 1966, pp. 7–8, June 15, 1967, p. 255.

15. Thomas O'Hanlon, "The Unchecked Power of the Building Trades," *Fortune*, December 1968, p. 102; *Wall Street Journal*, October 4, 1966.

16. M. R. Lefkoe, *The Crisis in Construction: There Is an Answer* (Washington, D.C.: Bureau of National Affairs, 1970), p. 5.

17. *Engineering News-Record*, May 8, 1969, p. 61, May 15, 1969, p. 67.

18. *Ibid.*, April 24, 1969, p. 64.

19. Lefkoe, *Crisis in Construction*, pp. 106, 147.

20. *Engineering News-Record*, November 28, 1968.

21. "Coming to Grips with Some Major Problems in the Construction Industry" (New York: Business Roundtable, 1974), pp. 16, 19.

22. *Ibid.*, p. 52.

23. *Engineering News-Record*, December 16, 1982, p. 132.

24. *Ibid.*, December 13, 1984, p. 58, January 13, 1983, p. 62.

CHAPTER 16

1. Secretary's Report to the 1970 Massachusetts State Carpenters Convention; Report of Second District board member to the 1970 Convention, in author's possession.

2. *Engineering News-Record*, July 6, 1972, p. 27.

3. *Ibid.*, April 4, 1968, p. 23.

4. 1964 state convention proceedings, p. 19.

5. 1968 state convention proceedings, p. 12.

6. *Engineering News-Record*, November 22, 1973, p. 43.

7. *Ibid.*, January 22, 1976, pp. 135–40.

8. *Ibid.*, January 29, 1976, p. 47; U.S. Bureau of the Census, *Census of Construction Industries, 1977, Area Series, New England States* (Washington, D.C.: Government Printing Office, 1981); Gerald Moody, "Regional Differences in Construction Activity," *Construction Review*, March–April, 1982, p. 7.

9. Reports of the Executive Board to the 1973 state convention; 1977 state convention proceedings, p. 18.

10. Secretary's Report to the 1973 state convention.

11. See Herbert Northrup and Howard Foster, *Open Shop Construction* (Philadelphia: Industrial Research Unit, Wharton School, University of Pennsylvania, 1975).

12. *Engineering News-Record*, April 27, 1972, p. 43, October 26, 1972, p. 96, May 15, 1975, p. 38.

13. Martin Seppala to Joseph Muka, letter in author's possession, January 6, 1967.

14. *Patriot–Ledger* (Quincy), July 22, 23, 28, August 28, 1976.

15. Secretary's Report to the 1975 state convention; 1981 convention proceedings, pp. 64, 77; 1983 convention proceedings, p. 229.

16. *Engineering News–Record*, November 11, 1982, p. 5, May 3, 1984, p. 28, November 8, 1984, p. 68.

17. *Ibid.*, January 10, 1985, p. 10, January 9, 1986, p. 52, June 14, 1984, p. 56.

18. John Avault, "Boston's Redevelopment: Economic, Fiscal, and Neighborhood Impacts; Private Investment Projects Completed, 1975–82, Scheduled 1983–1986, and Planned 1987 and Later," Boston Redevelopment Authority, Research Department, June 1984.

19. *Ibid.*

20. 1984 state convention, p. 202.

21. *Boston Globe*, October 23, 1984.

22. Stephen Tocco, "Power Grab by the Unions," *Ibid.*, August 2, 1983.

23. 1982 state convention, p. 9; 1973 state convention, p. 73.

24. 1981 state convention, p. 50.

25. *Engineering News-Record*, January 13, 1983, p. 62.

26. *Ibid.*, November 24, 1983, p. 62; *Carpenter*, February 1912, p. 27.

27. 1982 state convention, p. 9.

28. *Ibid.*, "Address by Anthony Ramos to the Southwestern Labor Studies Association Conference, April 29, 1983," *New Labor Review* 6 (Spring 1984), p. 8.

29. 1973 state convention, p. 7; *The Labor Page* (Boston), June–July, 1983.

CHAPTER 17

1. Proceedings, 1982 Massachusetts State Carpenters Convention, p. 115.

2. Steven Allen, *Unionized Construction Work-*

ers Are More Productive (Washington, D.C.: Center to Protect Workers' Rights, 1979); *Engineering News-Record*, August 25, 1983, p. 139. See Herbert Northrup, *Open Shop Construction Revisited* (Philadelphia: Industrial Research Unit, Wharton School, University of Pennsylvania, 1984), pp. 53–56; *Engineering News-Record*, June 29, 1978, p. 72; 1984 state convention proceedings, p. 148.

3. *Engineering News-Record*, February 14, 1985, p. 114.

4. Northrup, *Open Shop Construction Revisited*, p. 47.

5. 1980 state convention, p. 48.

6. *Engineering News-Record*, November 11, 1971, p. 88.

7. "Personnel: Finding the Right Team," *Professional Builder*, July 1982.

8. National Association of Homebuilders, *Economic News Notes* (Washington, D.C.), May 1983, p. 3.

9. "Builders: Residential and Commercial," *Wall Street Transcript*, August 2, 1982.

10. Robert Simison, "Mass–Output Methods Help Fox & Jacobs Gain Leadership in Housing," *Wall Street Journal*, March 29, 1978.

11. David Montgomery, "Beep, Beep, Yale's Cheap: Looking at the Yale Strike," *Radical America* 18, no. 5 (September–October 1984), p. 15.

CHAPTER 18

1. *Engineering News-Record*, October 16, 1969, p. 29.

2. UBCJA *Proceedings*, 1984.

3. John Daniels, "Industrial Conditions among Negro Men: Boston," *Charities* 15, no. 1 (October 7, 1905), pp. 35, 37, 38; Seaton Wesley Manning, "Negro Trade Unionists in Boston," *Social Forces* 17, no. 2 (December 1938), p. 261.

4. Manning, "Negro Trade Unionists," p. 258.

5. U.S. Commission on Civil Rights, Massachusetts Advisory Committee, *Contract Compliance and Equal Employment Opportunity in the Construction Industry* (Washington, D.C.: Government Printing Office, 1969), p. 433.

6. 1965 Massachusetts State Carpenters Convention Proceedings, President's Report.

7. Manning, "Negro Trade Unionists," p. 261;

U.S. Commission on Civil Rights, *Contract Compliance*, pp. 284, 62, 195.

8. Dennis Derryck, *The Construction Industry: A Black Perspective* (Washington, D.C.: Joint Center for Political Studies, 1972), pp. 28, 29; U. S. Commission on Civil Rights, *Contract Compliance*, pp. 234, 247.

9. *Engineering News-Record*, July 11, 1974, p. 17.

10. Harvey Lipman, "The Clearing House Builds," *Boston Phoenix*, January 17, 1978.

11. Rory O'Connor, "Hardhat Violence Coming to Boston," *ibid.*, June 23, 1976.

12. Barbara Lipski, "Minority Participation in the Building Trades," unpublished paper, 1984, pp. 10–11.

13. *Ibid.*, table 2, table 5.

14. Gary McMillan, "In Craft Unions, Brotherhood Not for All," *Boston Globe*, April 27, 1983.

15. These figures on apprenticeship are from the Massachusetts Department of Labor and Industries, Division of Apprenticeship Training.

16. *Engineering News-Record*, March 29, 1979, p. 27.

17. *Ibid.*, August 25, 1977, p. 8, March 29, 1979, p. 27.

18. Figures come from the Massachusetts Department of Labor and Industries, Division of Apprenticeship Training.

CHAPTER 19

1. Frederick W. Taylor and Sanford E. Thompson, *Concrete Costs* (New York: John Wiley, 1912), p. 94, 94n.

2. *Ibid.*, pp. 665–66, 669–70, 57.

3. Harry Braverman, *Labor and Monopoly Capital* (New York: Monthly Review Press, 1974), p. 90.

4. Taylor and Thompson, *Concrete Costs*, p. 86.

5. *Ibid.*, pp. 93, 104, 479.

6. Robert Hoxie, *Scientific Management and Labor* (New York: A. M. Kelley, 1966), pp. 132, 131.

7. UBCJA Constitution, 1888, p. 30.

8 Solomon Blum, "Trade-Union Rules in the Building Trades," in *Studies in American Trade Unionism*, ed. Jacob Hollander and George Barnett (New York: Henry Holt, 1912), p. 302.

9. *Carpenter*, January 1882.

10. William Haber, *Industrial Relations in the*

Building Industry (Cambridge: Harvard University Press, 1930), p. 226.

11. "The Building Situation in Boston," Hearings before a committee of the Boston Chamber of Commerce (Boston: Boston Chamber of Commerce, 1921), p. 620.

12. Advertising copy prepared by Brown & Root Construction Co. for national publication, 1981, in author's possession.

13. Taylor and Thompson, *Concrete Costs*, p. 479.

Index

In-text photo references are in italic type.

Amalgamated Society of Carpenters and Joiners, 42, 76
American Federation of Labor (AFL–CIO), 43, 85, 92, 110–11, 122–23, 160, 163, 225
American Plan, 105–9, 222
Anctil, Arthur, 115, 133, 134, 150, 151
Apprenticeship: in colonial era, 21–22; current state of, 170; employer attitudes toward, 51–52; as entry to trade, 2; as mechanism to control labor supply, 18–19, 51; nature of, 3–4; union vs. nonunion, 190, 215
Arkil, Nazadeen, *216*, 216–17
Associated Builders and Contractors (ABC): growth of, 161, 162–63, 167; and minority hiring, 215; and organization of work, 188–90, 222, 224; public image of, 167–68
Associated General Contractors (AGC): and apprenticeship, 2; and minority hiring, 211; and negotiating, 158; and 1956 strike, 153; and open shop, 163
Audley, James, 2, 4

Bathelt, Carl, 153, 166–67, 193
Bennett, Cliff, 156
Berkshire District Council of Carpenters, 56
Bernique, Leo, 1, 2, *3*, 4, 131, 148
Blomquist, Ellis, 2, 71, 114, 131, 164, 188, 192

Book of prices, 22
Boraks, Gordon, 171, *192*, 192–93, 194
Boston Building Trades Council: Boston carpenters and, 89, 112; and cooperatives, 92; during Depression, 117, 124; and Eliot controversy, 83; and WW I strike, 86–87
Boston Building Trades Employers Association (BTEA): antiunion attitudes of, 103–4; during Depression, 117–18; and 1919 strike, 89; and 1921 strike, 106–9; and 1956 strike, 153
Boston Carpenters Promotional Education Program (BCPEP), 170
Boston District Council of Carpenters, 68, 79, 83, 154, 166, 168–70
Boston Jobs Policy, 213–14
Bowen, John, 115–16
Bruno, Angelo, 1, 5, 135, *135*, 162, 167, 193–94, 224
Bryant, Bob, xvi, 2, 166, 193
Building cooperatives: in Boston, 91–94; in Mass., 92; in U.S., 92
Building Trades Unions Construction and Housing Council, 91–94
Burns, John, 161, 163
Business agent: contemporary, 169, 195–97; during Depression, 115, 119, 151; origins of, 78–84; and union rules, 73
Business Roundtable, 159–60, 161, 163, 222

Calhoun, Faith, 218
Campanelli Co., 134–35, 164
Cannon, Omar, 211, 213–14
Carabetta, Joe, 164
Carpenter, 19, 40, 47, 79, 85, 90, 114, 209
Carpenters Builders Association (CBA), 44–46
Christie, Robert, 24
Clinkard, Joseph, 44, 45, 79, 82
Closed shop, 50–51, 55, 58, 90, 109
Coffee break, 155
Cogill, John, 79, 172
Company man, 150
Concrete formwork, 145–46, 193–94
Construction Industry Stabilization Committee (CISC), 161–62
Conte, Phil, 154
Cooperatives. *See* Building cooperatives
Corbett, Joe, 209, 210, 216
Coulombe, Leo, 1, 4, 113, 114, 132, 146, 150, 151, 223–24
Craft unionism, 18–19, 225–26
Croteau, Richard, 2, 4, *5*, 114, 132, 133, 135, 149, 151, 154, 172, 173, 196, 223
Curley, Mayor James Michael, 109, 118, 155

DeCarlo, Angelo, 115, 147, 150–51
Double-breasting, 165, 168, 225
Drywall, 71, 147–48, 192
Dual wage system, 135–36
Duffy, Frank, 19, 76, 85

Edwards, George, 91–94
Eidlitz, Otto, 52
Eight-hour day, 41–47, 49–50
Eliot, Charles, 82–83
Emanuello, Joseph, 2, 133, 134, 135, 146, 147, 150
Ethnic local unions, 70–71
Evers, Tom, 215

Fletcher, Leo, 211, *211*, 212, 213
Foremen, 104, 222
Foster, Frank, 48, 72, 74, 80
French-Canadian carpenters, 69–70, 148

Gallagher, Ed, 134, 149, 151
Gompers, Samuel, 43, 48, 77, 85–86, 110, 123, 197
Gow, Charles, 104, 105, 106
Grant, Luke, 80
Green, William, 110, 123
Greenland, John, xiv, 119, 147, 153, 163, 215
Griffin, Dick, 171, 172
Grover, Reginald, xiii, 51
Gunning, Tom, 163, 189

Harrington, Tom, 1, 4, 73, 114, 132, 151, 154–55, *155*, 188, 193, 224–25
Health and welfare fund, 154
Henley, Ed, 114, 115
Holyoke Building Trades Council, 88, 120
Holyoke District Council of Carpenters, 115, 120
Hopkins, Harry, 126–27, 129
Housebuilding, 17, 132, 133–36
Hoxie, Wilbur, 132
Huber, William, 76, 209
Huddell, Arthur, 4, 87, 105, 106
Humphrey, Harold, 5, 147–48, *149*
Hutcheson, William, 86, 87, 123–24

Jones, Sharon, 216, *217*, 218
Joslin, Arthur, 104
Journeymen, 18, 21–23
Jubenville, Bob, 132–33, 145–46
Jurisdictional disputes, 75–77, 162

Kronish, Richard, 169

LaFrancis, W. J., 73, 82
Landry, Ernest, 1, 2, 69, 114, 132, 146, 148, 149–50

Leighton, Joel, 2
Leitao, Joseph, 112, 130
Lia, Joseph, 133, 189, 194
Lloyd, Harry, 43, 48, 68
Luther, Seth, 23
Lynch, James, 78, 80

MacKinnon, John, 1, 71, *72*, 73, 114, 115–16, 124, 133
Manufactured housing, 160–70
Marshall, Bob, 151, 168, 173, 188, 190, 196, 197, 215
Marshall Carpenters Training Center, 170, 188
Martin, Jim, xvi
Massachusetts Development Finance Foundation, 169
Massachusetts State Council of Carpenters, 126, 161
Master Builders Association (MBA): antiunion attitudes of, 51, 222; and eight-hour strikes, 41–44; and 1921 strike, 106
Master carpenters, 21–23, 25
Master Carpenters Association (MCA): 49, 51, 74
McGuire, Peter: and eight-hour campaigns, 43, 46–47; as founder of UBCJA, 40; on jurisdictional disputes, 75–76; on Panic of 1893, 49, 54; and vision of unionism, 67–68, 222
McNeill, George, 42, 44, 47
Molinari, Mike, xvi
Mroz, Mitchel, 3, *3*, 5, 6, 132–33, 135–36, 145, 156–57, 167, 170

Nason, John, 91–94
National Association of Builders (NAB), 47, 53, 54
National War Labor Board, 87
New Bedford Building Trades Council, 88
Nicmanis, Eric, 134
Norcross Brothers, 42, 44, 46, 52, 76–77

Open shop: 1921–22 campaign for, 108–9, 110, 111; in post–WW II housebuilding, 134–35; rise of, in 1970s and 1980s, 163–73, 188–92; strikes over, 1900–1916, 54, 56; work organization policies of, 222–23, 224

Operation Turnaround, 172
Organizing, 171–73
Osborn, Arthur, 215
Overtime pay, 51, 104, 109, 120, 155, 159

Pension fund, 154, 168–69
Peterson, Enock, 114, 119, 124, 133, 224
Petitpas, Joseph, 1, 5, *6*, 150, 224
Phalen, Thomas, 3, 6, 71, *72*, 114, 131, 133, 150, 154
Piecework, 22, 25, 78, 134, 191, 192, 221
Pile driving, 4, 193, 224
Potts, John, 68, 82, 83
Power, Joe, 173, 188, 189, *189*
Pratt, Oscar, 72, 114–15, 131, 146, 148, 151, 154, 196
Prudential Center, 155–56, 224–25
Public works: as Depression employment, 122–28; union attitudes toward, 126–29

Rickard, Harold, 2, 115, 119
Rickard, Tom, 1, 4, 5, 114
Routen, Bob, 6

Safety, 5–6
Sayward, William, 42, 47, 53, 54
Seasonality, 4, 133
Self-employment, 17
Seppala & Aho, 164–65
Sewell, Chester, 6, 114, 134, 135, 146
Shea, Donald, xvi
Sheetrock. *See* Drywall
Shields, William, 45, 68
Short, John, 208
Silins, Andy, xvi, 170
Smith Charities, 2
Specialization, 149, 191–92, 194, 223, 225
Springfield District Council of Carpenters, 55, 68, 73, 114, 115, 116, 118, 120, 126, 147, 154
Starrett, William, 16
Stewards, 73
Stewart, Ralph, 104, 105
Sympathy strike, 57, 74–75, 77, 112

Taylor, Frederick W., 219–21, 222, 223

Technological innovation, 24, 137, 146–47, 225
Ten-hour day, 23, 41
Third World Jobs Clearing House, 212, 213
Thomas, Bob, 1, 2, 145, 146, 148–49, *149*, 151, 156
Thompson, Sanford, 219–20, 223
Tocco, Stephen, 167–68, 188–89, 222, 224
Tramping, 4, 48
Turner, Chuck, 210, 212, 213, *214*, 215

Uniform wage, 19, 221, 223, 226
United Brotherhood of Carpenters and Joiners of America (UBCJA): in craft vs. industrial unionism debate, 18; decline of in 1920s, 111, 113; decline of, in late 1970s, 156, 165; and eight-hour campaigns, 43; founding of, 40; future of, 225; in jurisdictional conflicts, 75–76; and minorities, 208–9, 212; in Panic of 1893, 48–49; reasons for rapid growth of, 52–54, 67
UBCJA locals:
Local 23, 56

Local 33, 40, 41, 42, 44, 45, 48, 67, 79, 156
Local 40, 208
Local 56, 49, 154, 225
Local 67, 71
Local 93, 46
Local 96, 70
Local 108, 41, 49, 68, 166
Local 157, 71, 113
Local 183, 79
Local 192, 41
Local 275, 134
Local 402, 167
Local 417, 41
Local 444, 57, 114
Local 455, 40
Local 549, 83, 119
Local 624, 50
Local 762, 79
Local 847, 116
Local 877, 56
Local 1006, 111
Local 1105, 117
Local 1416, 131
United Building Trades Council (UBTC), 89–90, 106–8
United Carpenters Council, 51, 74
United Community Construction Workers (UCCW), 211

Valli, Al, 2, 115, 119, 151
Voluntarism, 123

Wage rates: in colonial era, 21; in 1832, 23; in late 1800s, 24; in 1901, 50; in 1902, 51; in 1904, 54–55; in 1906, 56; in 1919–23, 106, 108, 109; during Depression, 116–17, 118, 127; in 1950s, 132, 134, 153; in 1969–71, 209; in 1981–86, 166; in nonunion sector, 134, 189
Walking delegate. *See* Business agent
Walsh, Barney, 148, 151, 169, 170, 171, 196, 212, 215
Weatherbee, Bob, 3, 155, 172–73, 188, 191, 193, 195
Weiner, Paul, 72–73, 115, 130, 131, 146, 224
Weiner, Manny, 113, 114, 131
Weinstein, Michael, 192, *192*, 223
Williams, Mary Ann, 217, *217*
Women in Construction Project, 216
Worchester District Council of Carpenters, 92, 111, 114, 119
Works Progress Administration (WPA), 122, 126–29, 130
Work-sharing, 119–20
Wright, Carroll, 70